Advances in SHRIMP AQUACULTURE MANAGEMENT

ABOUT THE AUTHOR

Dr. S. Felix is the Professor of Department of Aquaculture at Fisheries College and Research Institute (Tamil Nadu Veterinary and Animal Sciences University, Chennai), Tuticorin, Tamilnadu. He has over 27 years of experience in shrimp aquaculture research. He had his training abroad on hi-tech-biosecured shrimp farming systems in USA as an FAO Fellow. The raceway facility he has prototyped and demonstrated for shrimp farming was the first of its kind for India.

The 'Maritech Research Centre' established by him at Tuticorin represents a number of state-of-art facilities for innovative mariculture activities. He has published 4 books, 12 manuals and over 100 research papers / reports, mostly on shrimp aquaculture research.

The technologies developed by him through funded projects (DBT, ICAR, NADP, etc.) include MSCD feed for shrimp larvae, raceways for shrimp brood and nursery management, LDPE lined growouts, proPO for shrimp immunostimulants, etc. The countries he has visited include USA and Australia.

Advances in
SHRIMP AQUACULTURE MANAGEMENT

S. Felix
Professor,
Fisheries Biotechnology Centre,
Fisheries College and Research Institute,
Thoothukudi – 628 008,
Tamil Nadu, India

2024

Daya Publishing House®

A Division of

Astral International Pvt. Ltd.

New Delhi – 110 002

First Published, 2009

ISBN: 978-81-7035-547-2 (HB)

Publisher's Note:

Every possible effort has been made to ensure that the information contained in this book is accurate at the time of going to press, and the publisher and author cannot accept responsibility for any errors or omissions, however caused. No responsibility for loss or damage occasioned to any person acting, or refraining from action, as a result of the material in this publication can be accepted by the editor, the publisher or the author. The Publisher is not associated with any product or vendor mentioned in the book. The contents of this work are intended to further general scientific research, understanding and discussion only. Readers should consult with a specialist where appropriate.

Every effort has been made to trace the owners of copyright material used in this book, if any. The author and the publisher will be grateful for any omission brought to their notice for acknowledgement in the future editions of the book.

Published by : **Daya Publishing House®**
A Division of
Astral International Pvt. Ltd.
– ISO 9001:2015 Certified Company –
4736/23, Ansari Road, Darya Ganj
New Delhi-110 002
Ph. 011-43549197, 23278134
E-mail: info@astralint.com
Website: www.astralint.com

Digitally Printed at : **Replika Press Pvt. Ltd.**

Dedicated to . . .

Dr. M.DEVARAJ

Ex-Dean of FC&RI, Tuticorin
Ex-Director of CMFRI, Kochi

☆ A silent reformer !

☆ A committed Scientist !

☆ An industrious individual !

☆ An administrator with vision for Fisheries Development !

☆ A teacher of exemplary teaching skill !

☆ A personality of down to earth simplicity !

☆ A technologist with the human touch !

☆ A great inspiring force in my carrier !

– S. Felix

Foreword

Shrimp aquaculture is still a significant economic activity in India although its rapid expansion has been curbed by judicial and environmental hurdles. There is no other aquaculture group which has attracted as much technological sophistication and investment as the shrimp and prawn. This is ample testimony to the potential and viability of shrimp aquaculture. Sustained growth of this sector calls for adoption of ecofriendly and biosecure farming practices at all levels including the hatchery, nursery and production stages as defined by the Aquaculture Authority of India.

This book written by Dr. S. Felix, Professor, Fisheries Biotechnology Centre of Fisheries College and Research Institute, Thoothukudi reflects the enormous experience and expertise of the author in shrimp aquaculture for over 25 years. The biosecure model for shrimp nursery practices he has introduced in India requires to be upscaled through their adoption by the industry. The chapters on nutrition, health care, bioremediation, water quality management and advanced farming systems deal with many vital areas of concerns in shrimp aquaculture.

I am sure that this book would immensely benefit the students, teachers, scientists, farmers, and the industry.

M.DEVARAJ

Dr. M. Devaraj, *Ph.D.*
Ex-Director, CMFRI

Preface

Shrimp aquaculture since its introduction in India as a commercial venture 25 years back has contributed immensely to the development of aquaculture sector in our country in general and export market promotion in particular. Despite its phenomenal growth, its true that the sector is yet to seen its peak performance. The fact remains that India is nowhere near China, the No. 1 ranked country today in farmed shrimp production. This was due to the fact that shrimp farmers in our country are still following their own practices mainly through seeing and learning from their neighboring farms. Farmers learn on trial and error basis and always try to correct the technical errors by themselves. There is hardy a institute-industry linkage or institute-industry-line department interactions / dialogues existing in our country, neither to address the problems nor for seeking remedies.

Unfortunately we have with us, industrious but at the same time argumentative farming community, talented but self-doubting scientific fraternity and to cap-it-all an exploiting commercial business houses all together have driven shrimp farming in India to a point, beyond that it is not able switch gears for its growth.

Shrimp farming has been witnessing tremendous changes in other countries, including smaller and economically weaker (compared to India) nations. Transformation and technical boom such as SPF & SPR stocks, biosecurity, indoor shrimp houses, zerowater concept, green house mechanism etc. are being introduced successfully. But we are not prepared to change a bit of our farming practices. 'Who is to bell the cat' to restore this 'industry of potential returns' from the clutches of 'business houses of only commercial interests' !.

A professional approach, with technical back-up, and management strategies could alone help this sector to come out of the present serious problems in which it is entangled.

This book could be a tool to bring about that realization in our shrimp farming sector, at the earliest as this sector has been proved to be the 'queen of all aqua farming practices' in the world.

This book deals with conventional and extensive farming practices, to the advanced farming practice such as raceway technology. It has been prepared particularly for the students who want to know the 'basics of shrimp farming'. The modern approaches in shrimp farming such as bioremediation, immunostimulants, probiotics, biosecurity, PCR in disease diagnosis, intensive culture technologies, etc. are dealt in detail. It's a comprehensive reading material covering wider aspects of shrimp farming and could be an information package for the students too.

Now that its over 30 years since the inception of Fisheries Colleges in our country, the professionals are started filling the strategic positions in various capacities including the farming sector. Its definitely going to change the scenario in the days to come. The experienced campaigners of this sector with the fisheries professionals should be able to raise this Golden Sector of Shrimp Farming in India, to newer heights of prosperity and to the point of highest economic returns.

S. Felix
E-mail: sfelix@sancharnet.in
sugafelix@yahoo.com

Contents

Foreword *vii*

Preface *ix*

Abbreviations Used *xv*

1. **Shrimp Culture Systems** 1
2. **Cultivable Penaeid Shrimp Species** 37
3. **Water Quality Management** 44
4. **Stocking Management** 57
5. **Shrimp Nutrition and Feed Management** 79
6. **Probiotics and Immunostimulants in Shrimp Aquaculture** 91
7. **Shrimp Raceways: Operation and Management** 104

8. **Diseases and Health Management** **131**

9. **Biotechnological Applications in Shrimp Aquaculture Management** **151**

10. **Scope for Genetic Improvements in Shrimp Aquaculture** **171**

Index **185**

Abbreviations Used

WSSV: White Spot Syndrome Virus

MBV: Monodon baculo virus

IHHNV: Infectious Hypodermal Haematopoietic Necrosis Virus

YHD: Yellow Head Disease

TSV: Taura Syndrome Virus

PCR: Polymerase Chain Reaction

RT-PCR: Reverse Trasncriptase PCR

YHDBV: Yellow Head Disease Baculo Virus

DNTPs: Deoxynucleotide triphosphates

BWSS: Bacterial White Spot Syndrome

TEM: Transmission Electron Microscopy

LDPE: Low density polyethylene

HDPE:	High density polyethylene
GSM:	Grams per square meter
PL:	Post larvae
UV:	Ultra Violet
RSF:	Rapid Sand Filter
PVC:	Polyvinyl chloride
DO:	Dissolved Oxygen
H2S:	Hydrogen sulphide
ANF:	Anti nutritional factor
IPNV:	Infectious pancreatic necrosis virus
SOD:	Superoxide disumtases
BMN:	Baculovirus midgut gland necrosis
GAV:	Gill Associated virus
BSD:	Bacterial shell disease
AADS:	Ascorbic acid deficiency syndrome
CMS:	Cramped muscle syndrome
CW:	Constructed Wetlands
RAS:	Recirculatory aquaculture system
SPF:	Specific pathogen free
SPR:	Specific pathogen resistant

Chapter 1
Shrimp Culture Systems

Different types of aquaculture systems are adopted for shrimp farming in India. The nature of site, species to be cultured, sustainability of source water, management practices, etc. make the shrimp farmers to decide the culture system to be selected. Generally three types of culture systems have been found to be ideal for undertaking shrimp farming. They are earthen ponds, lined earthen ponds and biosecured raceway systems.

The characteristics of these systems and their management practices are to be taken into account while selecting the site.

I. Criteria for Site Selection

(1) Earthen and Lined Earthen Ponds

The site suitability for constructing the farm for the above systems may be the same for both earthen and lined earthen ponds. The type and construction of shrimp ponds would largely depend on the type of management and species cultured.

A. Ecological Factors

(a) Water Quality

It includes all the physico-chemical and microbiological characteristics of the water. Certainly pH is an important aspect, and pH of water on, or adjacent to, the pond site should be within the range of 7.8 to 8.3. Water with a good growth of phytoplankton can usually be considered productive. If the water carries an excessive amount of sediment, a trap may have to be built into the water supply system. The amount of dissolved oxygen present near the bottom of the source of water to be used should be determined.

(b) Salinity

The normal salinity of water during high tide at different seasons of the year should be known. Especially important for rivers and canals is the subsurface intrusion of salt water under the freshwater. The depth of the top of the wedge at different tidal stages during normal weather should be ascertained. Also important is whether or not the tidal wedge persists during floods. The frequency of floods should be known. Just as important is the duration of freshwater conditions during flooding if the water source for the farm is a brackish water canal or creek.

(c) Tidal Characteristics

The tidal characteristics in relation to land elevation at the proposed site should be determined. This is critical to determine the tidal flow of the site which will decide the type of pumping to be used to fill the ponds, the elevation of the pond bottom, dike height, etc. In general, places where the tidal fluctuation is moderate, between 2 and 3 m, are most suitable for brackish water fish farms using the advantages of tidal flow to fill the ponds. Places where tidal fluctuations are more, over 4 m, are not suitable sites for pond construction because very large and expensive dikes would be required to prevent flooding during high tide. Also it would be more difficult to hold water in the ponds during low tides since due to the higher pressure, water loss and erosion from seepage, crab holes, etc., the seepage rate would be increased. Areas with slight tidal fluctuation, 1 m or less, are also not suitable for tidal ponds, because the ponds could not be filled or drained properly. So, if ponds are to be constructed in areas where the tide is less than 2 m or more than 3 m, the pump-fed-systems should be considered.

Actual measurements should be made at the pond site to determine the high and low tide benchmarks. One must keep in mind that tidal fluctuation is much less at certain times of the year. Tide tables should be consulted to determine these factors. Highest tides during past floods and storms also should be known. Sometimes the only way to acquire this information is from local residents. Wave action during normal tides, storms and monsoons should be known.

(d) Currents Prevailing in the Immediate Water Source Area

Knowledge of currents is important for planning erosion control measures to protect the dikes and the main gate as well as to determine the probability of sediment deposition in water control structures. Shifting mud or sand can block water supply canals or sluice gates, rendering effective water management impossible. As it is seldom practical to conduct surveys, one should ask local people if shifting sand or mud has ever been built up in areas near the pond site. Take into account the changing wind and current patterns at different times of the year.

(e) Rainfall

Important factors in the proposed site are maximum daily rainfall and annual distribution. The area of watershed and runoff in relation to the pond site should be looked into.

(f) Evaporation Rates

If evaporation is high, determine if there is an adequate supply of water to compensate the loss of pond water to maintain adequate water depth and proper salinity.

(g) Pollution

If the site is near a river, determine if harmful substances are used or released upstream. These would include such things as pesticides for agriculture and malaria control, mining wastes, industrial and urban wastes. Are these materials discharged continuously or only once in a while? Try and anticipate future pollution problems. Do not locate near a city that is growing rapidly or an area that is designed as a future industrial estate. Consult with the local government planning officials to investigate these aspects.

(h) Soil

In new areas where ponds are to be constructed for the first time, soil samples should be taken at ten random locations per hectare. Soil core samples may be taken at least to a depth of 0.5 m below the proposed pond bottom. This is because good soil might overlay unsuitable soil and a surface sample would not be sufficient.

Number of Soil Samples

In existing ponds, it is recommended that 12 samples be collected from ponds of 1 ha or less, and 25 samples from ponds of over 2 ha. In existing ponds, samples need to be taken only from the top 5 cm. A 100 ml portion of each soil sample should be placed in a plastic bucket to give one composite sample per pond. The composite sample should be mixed throughly. The soil samples are then taken to a soil-testing laboratory for analysis which include soil pH, clay content, etc.

Testing of Clay Content

Many coastal soils are high in peat or sand content and will not hold water. The potential pond soils must have a high and enough clay content to assure that the pond will hold water. A good field test to use in determining this aspect is to shape a handful of moist soil into a ball, if the ball remains intact and does not crumble after considerable handling, there is enough clay in the soil to provide a water tight seal. Sandy clay or sandy loam is the best for dike construction, because it is hard and does not crack when dried. Peaty soil is not a good dike material as it settles too much and may even burn when dried.

(i) Acidity and Potential Acidity

The effect of acid soils on brackish water production is generally attributed to low fertility of the soil, but acid soils may be the cause in many cases. Due to the conditions under which some coastal soils are formed, iron pyrite often accumulate. As long as these pyrite–containing soils remain submerged, they are subject to little change. When the land is drained to make fishponds the pyrites become oxidized producing sulfuric acid which cause the soil pH to become extremely low. Low soil pH can result in lowered pH of the pond water either by leaching from the pond bottom or by runoff of rainwater from the dikes during heavy storms.

The sulphuric acid formed when pyrite oxidizes not only affects pH of the pond water, it also affects soil minerals, releasing iron and aluminium which can bind up phosphates and other essential algal nutrients. This lowers the natural productivity of the pond and makes fertilization ineffective. The resulting lack of natural food causes slow growth of cultivable organisms.

Dikes made from acid sulfate soil develop vegetative cover very slowly and thus they are subjected to severe erosion. It requires added maintenance, both to rectify repairs on the dikes and to remove sediments from the pond. In addition, as the dikes are subject to oxidation, sulfuric acid and active aluminium and iron may be washed into the pond with eroded soil creating water quality problems.

When the pyrite containing soil becomes highly acidic after oxidation it is called "acid sulfate soil". The soil which will become acidic upon oxidation is called a "potential acid sulfate soil".

Acid sulfate soils can be identified easily by measuring their soil pH. The pH is 4.0 or less and mottles of the pale yellow mineral jarosite are usually abundant. In drained areas, an acid sulfate soil condition is characterized by a red colouration on the soil surface.

Potential acid sulfate soils are much more difficult to determine, because they do not become acidic until after oxidation. The soil can be acidified by exposure to air, but the extent and rate of the acidification process are regulated by chemotropic bacteria. Bacterial activity is low in dry soil, so it is best if the soil is kept moist. To do this, a soil sample is made into 1 cm thick cake and sealed in a thin plastic bag. The bag preserves the soil moisture and, if it is permeable enough to allow oxidation of the pyrite to proceed rapidly. The pH of the soil will be reduced to below 4.0 within one month if it is potential acid sulfate.

Considering the many problems associated with acid sulfate soils, a detailed soil survey is well advised before construction is started to develop shrimp culture ponds.

Construction in Areas with Acid Sulfate Soils

In areas with acid sulfate subsoil, special procedures are sometimes advisable in order to ensure pond fertility and prevent mortalities due to low pH. An economic feasibility study needs to be worked out for every farm to determine if such procedures are advisable. Some of the things which can be done are:

1. Excavate only enough soil from internal canals to construct the dikes and leave as much topsoil undisturbed as possible to serve as the pond bottom. In some cases, pumps might be required to fill this type of pond.
2. Scrap off the topsoil and set it aside. Then after the pond has been excavated, replace the topsoil over the pond bottom to cover the poor subsoil.

3. If the top layer of good topsoil is thick enough, alternate strips can be excavated twice as deep as necessary and then good soil from the unexcavated portion is placed in the deep portions to level the pond bottom.
4. Excavate the pond by sections. Excavate alternate 10 or 20 m widestrips within the pond and leave the strips between undisturbed. The pond is used for culture for several years, and after the excavated portions have aged, the remaining strips are excavated.
5. If the subsoil is not very acid, but has high potential acidity, a method of construction can be used in which the soil is kept moist so that it will not be oxidized and become highly acidic.
6. If acid soil, or potentially acid soil, has to be used for dike construction, the dikes bordering small nursery ponds can be surfaced with good topsoil. This is advantageous because acid runoff during rains has a greater effect in small ponds since the ratio of pond area to dike area is more critical.
7. If the dikes are to be constructed of acid soil, the nursery ponds should not be located next to large main dikes. This is recommended because runoff is much more from the large dikes than small ones, and as a result, more acid will be washed from them during rains.
8. If dikes are constructed of acid soils, the pond system should be designed to ensure that there is a minimum amount of seepage through the dikes into the ponds. This can be accomplished by having

the pond water inside the pond higher than water outside the pond. Drainage canals should be constructed around the pond to make sure that the water does not stand there.

9. A berm can be constructed near the water's edge to catch acid runoff during rains and prevent it from washing into the ponds.

(j) Percolation Rate

Knowledge of the rate of percolation of the soil will help in determining the extent of water loss through the pond bottom or dikes and can affect both design and management. If, for instance, a portion of the soil is good clay, it may be better to use this for the puddle trench and/ or centre core of the dike. If percolation can occur through the dike, and the dike soil is acid or potentially acid, it would be better to plan on having a positive water head in the pond to prevent acid from being washed into the pond by seepage through the dike.

(k) Depth of Topsoil and Characteristics of Subsoil

If the subsoil is unsuitable for the dikes, it may be better to construct the dikes of topsoil, or the poor subsoil can be used for the core of the dikes and the outer surface can be covered with topsoil. If the subsoil is highly acidic, it might be better to leave it undisturbed, reducing the amount of excavation, and filling the pond by pumping instead of tidal flow.

(l) Load Bearing Capacity

This is especially important if heavy equipment is to be used. It also helps to determine the number of pilings required under the water control gates, and the need for special foundations under the dikes.

B. Biological Factors

(a) Seed Resources

Determine if the seeds are available from hatcheries or dealers who obtain stock from the wild. If seeds are not going to be purchased from hatcheries (for those species for which hatcheries are yet to be developed), the local resources must be assessed to determine the species present and their seasonality of abundance.

(b) Predators, Competitors and Burrowing Organisms

The predominant pests vary from area to area and their presence in a given area may have an effect on management, construction or cost estimates. Now that most of the marine crustaceans are confirmed to be the carriers of white spot syndrome virus (WSSV), one needs to be cautious about these organisms.

(c) Wood Boring Organisms

Find out if these organisms are a problem in the area, if possible the extent of damage caused. The best way to determine what group causes the damage is to search out and examine old pieces of wood stuck in the ground, or the wooden boats of local people. This information can help to take the decision as to what type of material to use for sluice gate construction.

(d) Vegetation

The type of vegetation growing in the brackishwater area can be an indicator of productivity and soil type.

Following is a listing of some types of mangrove plants and the tidal zone they are usually associated with (Zinke, 1975).

Medium high tide: *Avicennia sp, Sonneratia sp, Excoecaria* sp. and *Thespesia* sp.

Normal tide: *Rhizophora* sp. and *Ceriops* sp.

Spring high tide: *Lumnitzera* sp. and *Acrostichum* sp.

Abnormal high tide: *Melaleuca* sp. and *Phoenix* sp.

Mangrove with growths of *Avicennia sp* have generally good soil and ponds built on them are generally productive. *Rhizophora* sp, *Bruguiera* sp, *Sonneratia acida* and most other mangroves with the same type of extensive, above ground, root system usually occur on acid soils which are less suitable for fish-ponds. Nipa and other trees with a high tanin content have a long lasting effect on ponds, causing low pH. Further, the number of trees and the size of their stems and root systems are also important factors in the cost of land clearing and excavation.

C. Social and Economic Factors

(a) Land Cost

Land cost need to be determined so that economic viability of the project can be evaluated.

(b) Accessibility

Accessibility is important for the transport of both construction equipments and materials, and for daily operations. Costs can increase significantly if materials have to be carried far by hand. If access to the pond site is by water, make sure that travel is possible even during the monsoon seasons.

(c) Availability of Labour

Local labour, that is residents living adjacent to the farm site, is the cheapest labour which can be obtained. Thus there will be a large saving in housing, transportation, food and other expenses. If workers are to be brought in from other areas they will have to be paid for these expenses. It is important to know the customs and tradition of the local people, as this will greatly affect the funding for labour. Identify the months when agriculture activities are in full swing. This will help in formulating programmes for repair of dikes and gates, stocking and harvesting. It may be difficult to get enough manpower during the time of intensive agricultural operations.

(d) Availability and Cost of Inputs

It is important to determine whether or not the supplies and equipment you need are available at the local area. Frequently the variety of inorganic fertilizer is greatly restricted and costs may be higher for non-agricultural use. Manures or other organic fertilizers might be difficult to obtain, or available in adequately requiring storage. If some materials such as aerators, probiotics, immunostimulants and drugs will have to be imported one should determine if there are any agents or extra costs involved.

(e) Marketing Outlets and Prices

This will have direct impact on management. If local buyers pay acceptable prices, the best form of management may be to practice partial harvesting, or to harvest one pond at a time, so that a small market is not flooded. If the shrimp have to be shipped some distance to a market, it might be better to plan to harvest and market large quantities at one time.

(f) Possible Legal and Institutional Constraints

This could include items such as: licensing requirements, land ownership laws, navigation laws, delays in processing applications, regulations against importing certain required materials (*i.e.*, machinery, equipment, etc.).

(g) Availability of Technical Assistance

This can be from government extension services, (line department) central government agencies or university research stations, or genuine private consultancy agencies.

(h) Users Competing for Land and Nearby Waters

The uses of nearby land and water should be assessed to determine what impact, if any, they will have on the project. Activities to be included would be such things as navigation, fishing, industry, public utilities, recreation and nursery areas. Problems can arise particularly if the activities of local people are disrupted. Make sure the project does not block a traditional right of way or interfere with work or recreational activities. It is recommended that plans for industrial development include provisions for rural districts as well as industrial districts so that the effects of industrial pollution on both agriculture and aquaculture will be minimized.

(2) Site Selection for Raceway Ponds

Though for the construction of earthen and lined ponds a vast area may be required, to construct raceway ponds, relatively a small area only is required. Further, the raceways need not require a site close to sea. Sub soil water with the salinity suitable for rearing shrimps *i.e.* 20–35 ppt also will be sufficient. Therefore the raceway ponds can be constructed even at interior places away from the shore. Further, the water requirement is also very less. While for the production of shrimp in earthen ponds it requires 40 tons/kg of shrimp production, in raceways the requirement is only 2 tons/kg of shrimp production.

II. Design and Construction

(1) Earthen Ponds

The physiological requirements and behaviour of shrimp are, in some cases, quite different from fish. By examining these factors, it should be possible to gain an insight into how to construct ponds suitable for shrimp farming.

General Layout

It is usually not a good practice to extend the pond area to the edge of source water along the sea or a major brackishwater river or canal. A buffer zone should be left for protection against erosion. Mangrove plants should not be cleared from these areas, and if not vegetation is growing, some should be planted.

Small streams, or other paths of runoff, should not be blocked unless a water diversion canal is constructed to carry off drainage water. Water which can not be drained from areas adjacent to ponds can sometimes seep into a pond in sufficient volume so that the pond bottom cannot be dried.

If it can be located elsewhere, the main water supply gate should not be located either at a bund adjacent to a river or facing the open sea. These areas are subject to strong currents and wave action which can cause damage to the gate which result in costly repairs.

It is important to place the sluice gate on the right spot so that the incoming water sets up a good circulation of water within a pond. This is best achieved by placing the gates near the corner on the short side of a pond. It is important, because letting in water is sometimes the only way to break up stratified pond water and prevent shrimp die-offs.

Areas with a high level of silt or mud in the incoming water will have serious problems. Ponds with internal canals along the dikes are ideal, because under this condition, the sediments are eventually deposited in the canals where they can be most easily removed. In an area facing a mud flat, it may be necessary to construct a small settling pond in which the silt can be trapped before the water enters the growing pond.

(a) Water Distribution System

Shrimp ponds need to have separate water intake and discharge canals. Water should be taken in at one corner of a pond and discharged from the diagonally opposite corner. This is especially important for large pond complexes with extensive canal systems. A single canal for intake and discharge of water from a pond complex has the following disadvantages:

1. All water drained from the ponds is usually not completely discharged from the canal and some of it will reenter the ponds the next time water is taken in.

2. The spread of disease from one pond to another is encouraged as water enters from one pond to another.
3. Water that is contaminated with hydrogen sulphide, ammonia or other contaminants can move from one pond to another.
4. If a single gate is used for both intake and discharging of water, exchange within the pond will be poor. Water at the far end is just moved toward the front during draining and then pushed back when new water is taken in.

Separate water intake and discharge canals in a pond complex have the following advantages:

1. Ponds can be filled better and water will not be contaminated by the discharge from other ponds.
2. The chance of spreading disease is reduced greatly.
3. A constant head can be maintained in the intake canal. This can reduce water loss through leaks in the pond dikes. It also cuts down seepage of water through the dikes and consequently reduces leaching of acid into the ponds from dikes with acid sulfate soils.
4. No conflict of usage should occur between farmers.
5. A better exchange of water is provided for individual ponds.
6. Flow-through systems can be used. With pumping, a constant head for continuous flow-through can be maintained.

(b) Dike Formation

The following specifications are recommended for dikes:

1. All low depressions should be filled in before dike construction is started (when the area is crossed by creeks or rivers, the portion of the dike running across these should be constructed first).
2. A 'puddle trench' is essential to prevent water seepage under the dike. The earth should be packed as it is replaced into the trench. The dimensions of the trench should be 0.5 to 1 m deep and 0.5 to 1 m wide depending on the size of the dike.
3. The following slopes are recommended for dikes built with good clay soil.
   - 2:1 when dike height is above 4 m and exposed to wave action.
   - 1:1 when dike height is less than 4 m and the tidal range is greater than 1 m.
   - 1:2 when the tidal range is 1 m or less, and the dike height is less than 1 m.
4. The crown should not be less than 0.5 m. The actual width depends upon the activities which will be performed during culture operations.
5. The main dike surrounding the farm should be 0.5 m above the highest tide or flood level recorded in the locality.
6. During construction 15 to 20 percent excess height should be allowed for shrinkage due to settling.
7. Construction should be in stages. First, build the dike to 1/3 of its final height all the way around the pond. Then build the height to 2/3, and finally

to its full height. This allows the base to consolidate so it can support the weight of the top portion.

8. A berm built on the inside of the dikes should be slightly above the water line. This will minimize the effect of wave action on the dikes. A berm is also an advantage when it is necessary to dig out and replace soil to repair damage caused by crab holes.
9. To calculate the cross-sectional area of a dike to determine the amount of soil needed, the following formula can be used:

$$\text{Area of cross-section} = \frac{(\text{Width of base} + \text{Width of crown})}{2} \times \text{Height}$$

The area of the cross-section of the dike is multiplied by the length of dike to determine the amount of soil required.

(c) Control of Erosion and Leakage

1. Growth of green cover on the completed dike should be encouraged to prevent erosion.
2. Mangroves planted at the water's edge will prevent erosion of the dikes. Mangroves can be therefore planted at the water's edge.

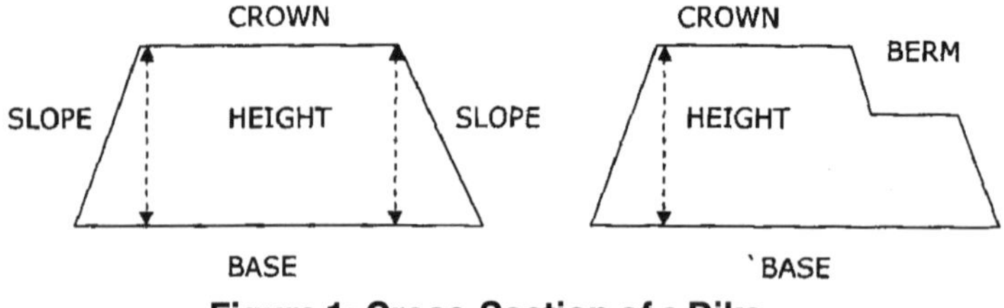

Figure 1: Cross-Section of a Dike

3. In large ponds small submerged dikes can be built 10-15 m from the dikes during construction. Wind waves will break up on these submerged dikes and the water control gates will not be damaged.
4. In areas where burrowing organisms are a known problem, damage caused by their burrowing can be prevented by incorporating a bamboo screen or plastic film in the puddle trench during construction.
5. The same materials can be placed in a cut made in the berm to repair damage. To stop leaks, slaked lime can also be added to a cut made in the berm. Easily applied as a powder it sets up hard on contact with water.

(d) Canals

The width of the canals would depend on the amount of water they need to carry. The following factors must be taken into account:

1. The volume of water which will be held in the ponds.
2. The time requirements for filling or draining the ponds.
3. The amount of rainfall which must be carried off in a given period of time.
4. Elevation of the canal bottom in relation to tide. For instance, tidal ponds cannot be drained during high tide, but a pond built at a higher elevation and filled by pumping could be drained during any tidal phase.
5. Other uses.

Canals which are to be used for harvesting should be 30 cm below the level of the pond bottom and the canals are of many types *viz.* diagonal, radiating and peripheral.

The canals formed in ponds help to maintain water depth; enhance circulation of water inside the pond; provide adequate hiding place to shrimps; to collect the silt coming with the incoming water and to harvest shrimps effectively.

(e) Water Control Gates

There are a number of types and sizes of water control gates and construction can be with many kinds of materials. There are, however, certain requirements that all control gates at a shrimp pond should fulfill.

1. A gate should first of all be of adequate capacity for the amount of water required to be taken in or drained.
2. It should be constructed in a position that water can be taken in and discharged at the bottom.
3. It must have provision for draining surface water from the pond.
4. The bottom of the gate must be at an elevation which permits all the water to be drained.
5. It should be water tight.
6. It must have slots or grooves for the placement of screens on the outside to keep trash undesirable species out of the pond, and on the inside to prevent shrimp from leaving the pond.
7. It should have a slot to install a net for harvesting during draining.

8. Must be durable.
9. Easy to operate.
10. Made up of locally available materials.

Antiseep boards or collars will prevent lateral seepage and resultant washouts. A rubber liner of automobile inner tube attached to the closure boards helps to make a good water tight seal. Winches can be used to remove boards. This allows the use of heavier, one or two piece boards. If gates designed for use in ponds with shallow water such as for lab-lab culture, the side boards can be placed inside the support bracing. This allows the boards to be replaced easily when they decay. Gates to be used in ponds with deeper water, such as for plankton culture, should have the side boards placed outside the support bracing. This is necessary because the greater pressure pushes the side boards inward and if the boards were inside the bracing, they would become loosened and water leaks would occur.

Construction of gates have to be supervised closely, especially with concrete gates. Otherwise, the sides might slope and different length boards would be needed for each level. Also in multilane gates, the widths might vary and the boards would not be interchangeable.

Concrete gates, if not properly constructed, would be better made of wood. If workmanship is poor, the gates might not hold water. If construction is faulty or the design is inadequate, repairs will be costly and there may still be no guarantee for safety. The following basic considerations should be taken into account when constructing gates:

1. Strong foundation is inevitable.

2. Adequate reinforcement against side way pressures from the dike and water.
3. The concrete is to be properly mixed and cured.
4. Adequate measures are taken to prevent under-cutting by seepage of water along the sides and bottom of the gate.

To make a strong foundation, bamboo stakes are driven into the ground to a depth of 1 m. The stakes should be 30 cm apart. The stakes are cut off, leaving a sufficient length to penetrate into the concrete slab.

If poured concrete is used, the mixture of cement to sand to gravel should be 1:2:5 (class B) for the wall and 1:2:4 (class A) for the flooring. If concrete hollow blocks are used, the mixture should be one part cement to seven parts sand. The amount of water added should be 22.2 litres per 45 kg bag of cement. Spacing of steel reinforcing bars should be 40 cm, centre to centre. Bars of 1.2 cm (1/2 inch) diameter are used for vertical reinforcement and 1 cm (3/8 inch) bars for horizontal. Both hollow blocks and poured concrete walls should be at least 15 cm thick. The proportion of cement to sand should be 1:3 in mortar for finishing. Mortar must not be applied more than 0.6 cm (1/4 inch) thick. The concrete should be cured (hydrated) for 21 days. This is done by covering the concrete with sacks and keeping the sacks moist for the whole 21 days period.

2. Lined Ponds

Earthen ponds are provided with lining on sides and bottom with the help of low density polyethylene (LDPE) or high density polyethylene (HDPE) sheets. The advantage is that the high level of water seepage from earthen ponds could be arrested by the lining. Salination of adjacent lands also could be prevented by lined ponds. Further, the pumping cost also is considerably reduced in such systems, as we need to fill up water only to compensate the water loss due to evaporation.

(a) Lining Materials

(*i*) LDPE Sheet

LDPE materials are available in different thickness ranging from 150 to 300 GSM (grams per square metre). They are costing less compared to HDPE sheets and they may lost for 5–6 years, if exposed to sunlight and more than 10 years if covered with clay/sand (Plate 1).

(*ii*) HDPE Sheet/Synthetic Fibre Spread

HDPE sheets/synthetic fibre sheets are available in different thickness ranging from 500–1000 GSM. Relatively costlier but long lasting, would lasts for more than 12 years even if it is exposed to sunlight (Plate 2).

Various other materials which are used for lining the ponds are:

1. Brick lining
2. Concrete lining
   a. Inset lining
   b. Slab type
3. Sheets
   a. PVC sheets
   b. Butyl rubber
   c. Poly propylene

Selection of Lining Materials

The inner lining material for the pond is selected based on the following factors:

1. Soil type
2. Cost of lining material
3. Cost of installation
4. Durability
5. Suitability

Method of sealing the polyethylene sheet for lining the shrimp grow out pond:

1. The surface of LDPE sheet to be used will be cleaned. It should be free from grease, dust and other dirt for better sealing.
2. Teflon pad will be used beneth the sealing surface for effective sealing (Plate 3).
3. According to the thickness of the LDPE/HDPE sheet the heat is regulated in the sealing machine.
4. To avoid damages due to excess heat a teflon sheet also will be kept over the sheet before sealing.
5. After sealing teflon sheet will be removed and wiped with a wet cloth.
6. Finally the effectiveness of sealing has to be checked (Plate 4).

(3) Raceway Ponds

Design and Construction of an Intensive Nursery Raceway System for Penaeid Shrimp Cultivation

The major problems facing the shrimp farming sector today is the poor survival and unpredictability of yield

Plate 1: Earthen Pond Lined with 250 gsm LDPE Sheet

Plate 2: Laying of 700 gsm Synthetic Fibre Sheet in a Raceways Tank

when juvenile shrimp are stocked into growout ponds. A management strategy has now been developed to increase juvenile shrimp survival and production predictability for

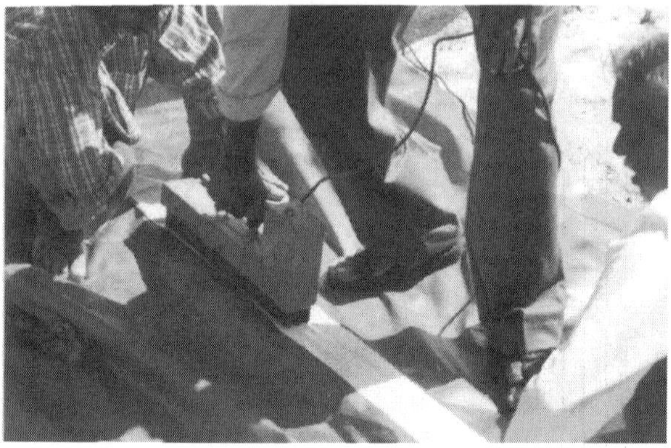

Plate 3: Heat Sealing of LDEP Sheets

Plate 4: Lining of a Pond

PL 15 to 30. By using raceways enclosed in greenhouses in temperate regions such as the southern U.S.A., Mexico, etc. the production period can be reduced in the grow out ponds. In other words, extension of the growing season by utilizing enclosed greenhouses as nurseries decreases the period required for the production of marketable–size shrimp in ponds. This "head starting" can result in production of two crops a year. For regions with a year-round growing season, the use of an intensive nursery management strategy ("head starting") will increase shrimp survival and predictability.

The technology described in this chapter are currently being used at the Texas A&M University facilities of the Texas Agricultural Experiment Station, Texas A&M University System, in Corpus Christi which has been adopted at Fisheries College and Research Institute, Thoothukudi with suitable modifications (Felix, 2004). The operating procedures of raceways described have resulted in high survival (above 90 per cent after 7 weeks) with a harvested biomass of about 1.9 kg/m^2, using stocking densities upto 3200 post larvae (PL) per square metre (Samocha, 2001). Nevertheless, the potential carrying capacity of this system is even higher. Densities of about 7800 PL per square metre have already been tested, with over 90 per cent survival for a similar growing period. Interestingly no stunting effect was observed when these shrimp were thinned to 220 and 300/m^2 and continued for another 50 days. Survival of over 80 per cent with an average weekly increase of 1.07 g were demonstrated.

This chapter provides a system description, with the significance of such components involved in the raceways based upon the design developed at FC & RI, Thoothukudi.

System Description

The description of nursery raceway system provided in this chapter is for the unit of two raceways.

A. Reservoir Pond

The reservoir pond is recommended for locations where significant changes in salinity are expected or if ambient water salinity is above 40 ppt. The reservoir allows adjusting salinity before pumping water to the raceways. A reservoir working volume of 100 m^3 is recommended for the two raceway unit with a total water volume of 120 m^3. To exclude large predators, reservoir intake pipes should be equipped with two filter bags of 420 and 800 µm screens. The reservoir can be designed to feed raceway pumps by gravity to avoid constant priming problems. An external swivel standpipe at the deep end is recommended for draining and reservoir overflow control. The Figure 2 provides a schematic view of the reservoir.

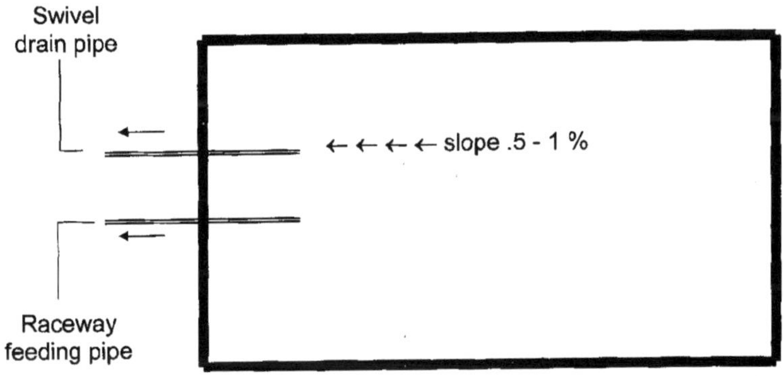

Figure 2: Schematic View of the Reservoir

Greenhouse Structure

The greenhouse facility currently in use at the Fisheries College and Research Institute, Thoothukudi has solid M.S. Iron framing with high seawater resistance. The roof and end walls are made of translucent corrugated fiberglass panels. The sidewalls are made of green shading (80 per cent) material. The structure is large enough to maintain two 45^3 capacity raceways with pumps and rapid sand filters. The greenhouse space is equipped with floodlight fixtures arranged in three rows. The structure

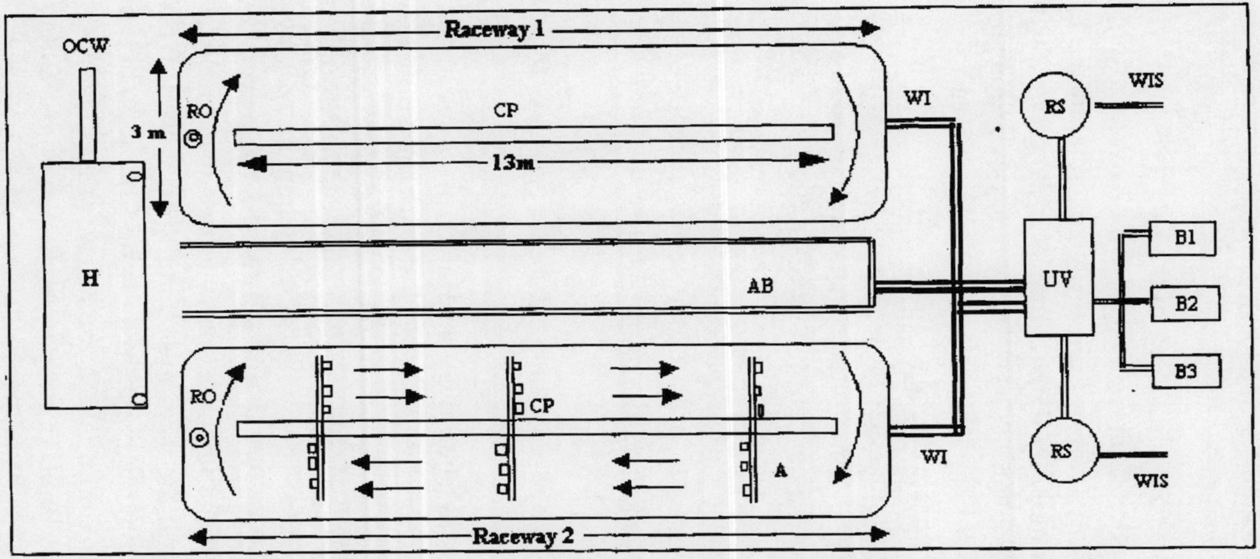

Figure 3: Raceways and Supporting Systems (Aerial View) (Felix, S., 2005)

A: Airlift system; AB: Aeration line from blowers; CP: Central partition wall (18 mm); B: Blower (5 HP each); H: Harvesting tank; RO: Raceway's water outlet connected to harvesting tank; OCW: Outlet to constructed wetlands (CWs); WI: Water inlet provision; WIS: Water inlet through Rapid Sand Filters; UV: Ultra violet sterilization of water; RS: Rapid Sand Filters (20000 LPH) can be constructed using the locally available, durable and cost effective materials.

Efforts are to be made to build the raceways with adequate elevation to allow juvenile shrimp to drain during harvest. Raceways can be built using concrete, fiberglass, plywood, or simply by lining a trench in the ground with high-density polyethylene (HDPE) membrane. Elongated raceways (30 m long; 3 m wide) with rounded end walls are adequate. Raceway walls can be vertical or with a 2:1 slope to the bottom. Each raceway will have a 25 m long centeral partition. A minimum water depth of 0.5 m and a bottom slope of 0.5 per cent with a 0.3 m deep sump at the deep end are suggested to improve waste removal (Figure 7). Freeboard of at least 0.15 m is needed for raceways. For raceways designed to produce larger-size shrimp, deeper freeboard or net cover will be needed to avoid shrimp losses due to jumping.

(ii) Pump, Pipe, Lines and Valve Layout for Raceway

Each raceway should be separately provided with a pump with a capacity of about 25 m^3/h. The pump will serve to optimize oxygen injection, enhance raceway water circulation, to filter water and to pump algae from one raceway to another. The pump is used to both bring in new water and recirculate water within the raceway. The raceway pump should have access to two water sources: the reservoir and the raceways. A multiport valve (M) can direct water from the pump to the sand filter (SF), bypass it, or send it to drain. Filling the raceways is done by the raceway supply pipe. This pipe is equipped with a 50 mm check valve and a 20 mm PVC ball valve. This valve can be

Plate 5: Raceways in Operation at FC&RI, Thoothukudi (Felix, S., 2005)

used temporarily for post larval acclimation. At the shallow end raceway, the pipe has three 50 mm PVC ball valves. One valve controls water flow into the raceway, the second valve controls water flow into the bottom manifold and the third valve feeds the common distribution pipe. This pipe is equipped with a bleed valve for flushing the pipe network.

Raceway water intake pipe also can be equipped with flow metre to facilitate daily water management. Installation of 50 mm freshwater outlets is recommended for areas where high ambient salinity is expected. An additional 20 mm freshwater supply pipe is needed for day to day routine cleaning.

(iii) Raceway Outlets and Filter Pipes

Each raceway drain outlet is positioned at the centre of the sump, half-way between the end of the partition and the raceway's end wall. The raceway water level can be controlled by a swivel external standpipe. Each raceway should be provided with a set of four filter pipes of the following screen sizes: 600, 800, 1000 and 2000 µm. Filter pipes are mounted on the outlet to avoid losing shrimp during water exchange. These are changed as the shrimps grow. Two types of filter pipes are needed: the all-perforated/slotted and the only-ends-perforated/slotted type.

(iv) Rapid sand Filters

Conventional swimming pool sand filter with manual backwash and filtration capacity of about 20 m^3/h is needed for each raceway. The sand filter can be used to filter the incoming seawater as well as the raceway water.

Plate 6: Raceways with Airlift Systems (Felix, S., 2005)

(v) Multiport Valves

The multiport valve is a multiposition valve with six operational modes: sand filter, backwash, rinse, circulation, waste and closed. The backwash and rinse modes serve to maintain the sand filter in optimal working condition. The circulation bypasses the sand filter. The waste mode can be used to drain the raceway without using the external standpipe. The closed mode is a safety position to avoid accidental drainage through the multiport valve when the raceway is not in operation.

(vi) Oxygen Injection

The liquid oxygen system includes a storage tank, pressure regulator, distribution pipe and valves, oxygen filters, oxygen flow meter, and oxygen diffusers. The distribution system can deliver oxygen into an oxygen

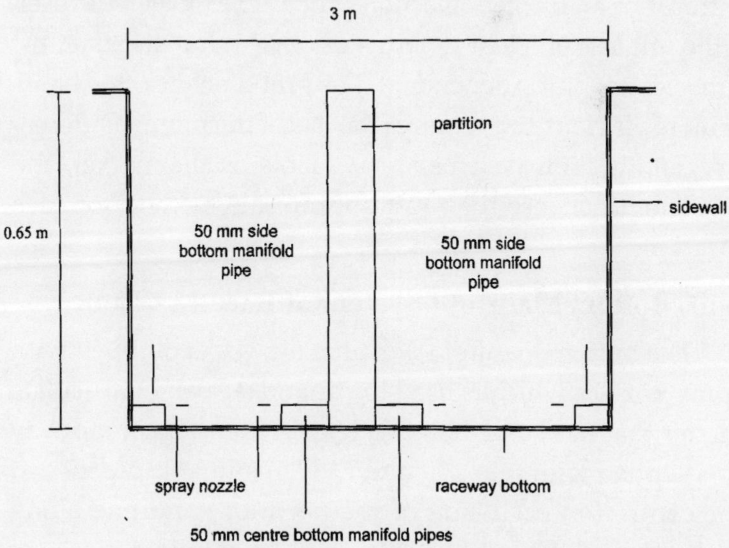

Figure 4: C.S. of Shrimp Raceway Tank

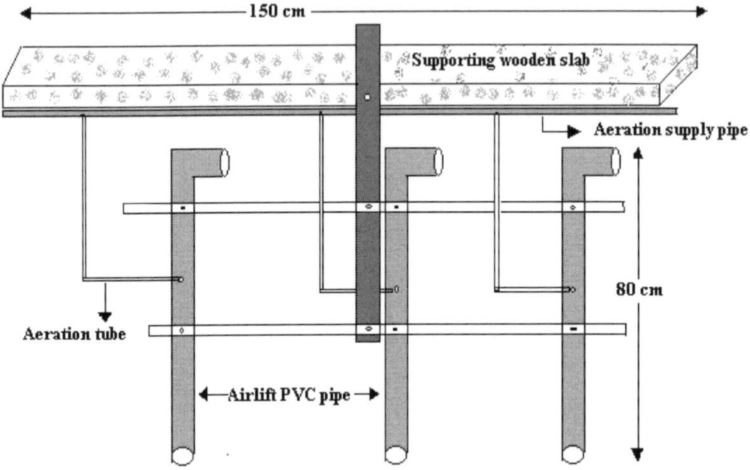

Figure 5: Longitudinal Section through Airlift and Structural Support (Felix S., 2005)

diffuser or an oxygen injection probe. The oxygen diffuser, although less efficient in terms of oxygen transfer than the probe, is not pump dependent. The probe, on the other hand, is more efficient, but requires the use of the pump. Different oxygen injection systems are available on the market, and one should choose the most appropriate method for each location.

(vii) Bottom Manifold System of Raceway Tank

The bottom manifold is a pipe network equipped with spray nozzles which is used to stir and suspend particulate matter that has settled on raceway bottoms. Each raceway is equipped with four 350 mm PVC manifold pipes, two at the centre (on each side of the partition) and one along each sidewall (Figure 8). All the bottom manifold pipes are interconnected for single valve operation. Manifold pipes are equipped with non-metallic, flattened-fan-shaped spray nozzles mounted in 45° PVC elbows.

(viii) Blowers and Air Distribution System

About 400 lpm of air will be needed for a nursery unit of three raceways. This air supply should be adequate for feeding a total of 50 mm airlift pumps, 1 m micropore air diffusers and 1000 litre *Artemia* hatching tanks. Air can be provided by any high-volume, low-pressure blower. A smaller unit should be maintained as a backup. A main air distribution pipe should feed six longitudinal pipes which provide air for the raceway. Each airlift pump and air diffuser is connected with a flexible hose to a valve.

(ix) Airlift Pumps

Airlift pumps are the vital component in a raceway. Water circulation in the raceway depends primarily on airlift pumps. In this device, air is introduced near the bottom of a vertical pipe which has a 90° PVC elbow at the upper section. The air, lifts the water in the pipe and sends it through the elbow. Air pressure, airlift pipe diameter, submergence depth, and the type of airlift pump being used are among the main factors affecting pumping rate. Use of the collar-type airlift pumps are generally recommended. In this type, air is forced through a ring of small-diameter holes on the footpiece. The airlift pumps are arranged in eight banks, four on each side of the partition, with three airlift pumps in each bank. The pumps are mounted on a structural support to allow suspension across the raceway. This support can be built from PVC, fiberglass or wood with required weights for adequate positioning of the airlift pumps.

(x) Air Diffusers

Each raceway is equipped with six 1 m-long air diffusers, made up of low-pressure, 25 mm rigid porous plastic pipe, providing air bubbles of 20 to 40 μm in size. Air is provided via two rigid PVC elbows to ensure continuous airflow. Air diffusers are attached to both sides of the partition just above the bottom manifold pipe.

(xi) Lighting Facility

The greenhouse is equipped with sufficient light fixtures to provide sufficient light for night feeding without disturbing shrimps and for helping with the night harvest. Lighting may also will be helpful during cloudy and rainy days to sustain algae in raceways.

(xii) Temperature Control

Building nursery facilities under greenhouses in temperate climates will help gain about 10°C above the ambient water temperature, depending the greenhouse efficiency, sunlight and water exchange. In places where the water temperature during early spring can drop occasionally below 10°C, space or water heaters are needed to maintain adequate water temperature during cold-weather periods. In tropical countries the greenhouse helps to maintain salinity and temperature in raceways. It protects the system from sunlight, dusty wind and rainfall and helps to maintain the temperature.

(xiii) Backup Generator

A backup generator with sufficient capacity to support the main raceway blower in case of power failure is indespensable. This generator should be equipped with an automatic switch for immediate start of blower in case of power failure.

Chapter 2

Cultivable Penaeid Shrimp Species

Many factors need to be considered while a farmer is deciding which species of shrimp is to be cultured. Due to its faster growth, large size and high price, *Penaeus monodon*, the tiger shrimp is generally considered the most desirable. However, as *P. monodon* is considered to be more susceptible for virus infections under tropical conditions alternate penaeid species can also be tried. It might be advantageous to minimize risk by culturing a species with a short growing period. It may be desirable to consider growing different species during different seasons depending on the variations of the environment or availability of seeds. The marketing potential of the species also has to be taken into account while selecting a species for farming.

Of the species of penaeid shrimp occurring within the region, the following are the ones most commonly cultured in India:

Penaeus monodon: Tiger shrimp

Fenneropenaeus indicus: Indian white shrimp

Penaeus merguiensis: Banana shrimp

Penaeus semisulcatus: Grass shrimp

There are few other penaeid shrimps which are cultured in other countries. They are *Penaeus japonicus* (Kuruma shrimp), *Litopenaeus vannamei* (American white shrimp), *Penaeus chinensis* (Chinese shrimp) etc. As an aid in choosing which species to culture, some advantages and disadvantages of the most common or well-known species are given below:

(i) *Penaeus monodon*

Advantages

(a) It attains a larger size. Shrimp with a size of 10 to 12 pieces/kg are not uncommon, and sizes of 5 to 7 pieces/kg have also been recorded.

(b) It is the fastest growing of all cultivable shrimps. In ponds, juveniles of 3 cm in length have been grown to a size of 75 to 100 g in only five to six months.

(c) Due to its large size, it brings a high price to the farmer. Over US$ 10 per kg of shrimp weighing 15 counts is always possible.

(d) It can tolerate a wide range of salinities *i.e.* 0 to 70 ppt. Salinity within the range of 10 to 25 ppt has appreciable positive effect on growth when food is sufficient. However, growth is reported to be slower at very low and very high salinities.

(e) It can tolerate temperatures upto 37.5°C. Mortalities are known to occur at temperatures below 12°C.

(f) It grows rapidly when fed either with animal or vegetable protein or the combination of both.

(g) Food conversion ratios (FCR) are favourable. Values as low as 1.2:1 have been reported using pellet feeds.

(h) It is hardy and not easily disturbed by handling.

(i) Hatchery technology is standardized.

(j) Withstands treatment and prophylactic measures undertaken in culture systems.

Disadvantages

(a) There is a sparse supply of quality wild brood stock for hatcheries.

(b) Seeds are usually expensive.

(c) Gravid females are difficult to obtain from the wild in sufficient numbers to support a large hatchery.

(d) It takes relatively a long growing period (more than 5 months) to reach the marketable size (40–50 g) which commands the best price. This increases risk of heavy losses from water quality deterioration, disease outbreak, etc.

(e) Difficult to harvest because it does not leave the pond with discharged water as readily as other species.

(f) The head to tail ratio is not as good as that of some other penaeid shrimp species. This could have an adverse effect on sales to the export market where only tails are desired.

(g) The exoskeleton is rather thick and processors find it harder to remove than those of most other species.

(h) Relatively more susceptible to white spot syndrome virus (WSSV) infection.

(i) Requires high protein feed, which naturally increases the cost of production. Water quality also deteriorates faster.

(ii) *Fenneropenaeus indicus*

F.indicus and *P.merguiensis* supposedly can be differentiated by five separate characteristics, but based on actual field surveys there are many individuals which do not have very clear distinction on these five points.

In spite of the taxonomic resemblances, there are indications of behavioural differences between the two species. In the Philippines, for instance, *F. indicus* is difficult to harvest by draining ponds, but in Thailand, *P. merguiensis* moves out of ponds readily when water is drained. In addition, *F. indicus* prefers sandy substrates and *P. merguiensis* is found most frequently on mud bottoms. However, the growth potential of these two penaeid shrimps would be more or less same.

Advantages

(a) They grow to a fairly large size and brings a good price.

(b) It is fairly fast growing, especially when young. At a density of 15 m^2, it can reach a size of 15–20 g in 16 weeks.

(c) Survival is high during the first three months of growth or upto a size of about 10 cm.

(d) Wild fry are usually abundant in estuarine regions where the adults are present.

(e) Gravid females are relatively easy to obtain from the wild in numbers sufficient to operate a hatchery.

(f) Females can be matured in captivity with relative ease.

(g) Shrimp moves out of a pond with water discharge, making harvesting easy.

(h) Good growth has been obtained in intensive culture with a feed having 35 percent protein, which is lower than that required for some other species.

(i) Hatcheries are available to supply seeds.

Disadvantages

(a) Relatively high salinity (above 30 ppt) is required for better growth. It has a wide tolerance to short-term exposure to salinity extremes, but dies with long exposures.

(b) Mortalities occur at temperatures above 38°C.

(c) There is a significant size difference between sexes.

(d) It can not with stand rough handling as either a juvenile or an adult. Juveniles are weaker than those of *P. monodon* during transport.

(e) Wild fry are more difficult to identify than most other species of *Penaeus* or *Metapenaeus*.

(f) Difficulties encountered in culturing this shrimp for more than 4 months without heavy mortalities.

(iii) *Penaeus semisulcatus*

Advantages

(a) Artificial propagation of larvae is relatively easy.

(b) Shrimp reaches a large size and brings a good price.

Disadvantages

(a) This species has not been successfully cultured to marketable size despite numerous attempts.

(b) In ponds, growth is slow and mortality has been high.

(c) It requires relatively high salinity water.

(iv) *Litopenaeus vannamei*

Advantages

(a) They are highly tolerant to low temperature ranges.

(b) They require a relatively short growing period, only three months, to attain marketable size.

(c) Hatchery technology is available.

(d) Survival in ponds is usually high.

(e) Harvesting can be accomplished easily by catching them as water is drained from a pond.

(f) Only species for which disease resistance (genetically manipulated) seeds are available (SPR).

(g) Virus free brood stock availability is ensured.

(h) They are easy to mature in captivity and larval culture is relatively easy.

(i) Tolerance to handling is more.

(j) Well-suited for raceway based intensive farming.

(k) Market value is very high due to the consumer preference prevailing for this species in USA and European markets.

(l) Low protein feeds (25–35 per cent) are sufficient in grow out systems.

(m) Tolerates high stocking densities in grow outs and raceway ponds.

(n) Individual weight as high as 30 g have been achieved in Indonesia in growout ponds, adopting unique culture techniques.

Disadvantages

(a) They do not generally grow to a large size.

(b) Susceptible for Taura Syndrome Virus (TSV) infection.

Chapter 3
Water Quality Management

The emphasis on water quality standards for shrimp aquaculture is much more than the other farming systems. Shrimps are considered to be more sensitive to abrupt changes in water quality parameters and hence utmost care needs to be exercised to maintain the water quality in shrimp farms.

(*i*) Temperature

Both growth and survival are directly influenced by temperature variations. Generally, the rate of growth increases with temperature, but at higher temperatures mortality also increases. While each species has its own optimum temperature range, temperatures between 26 and 30°C are generally considered suitable in terms of maximum yield. That is, growth is relatively fast and survival is high.

However, temperatures above 32°C should be the cause for concern. The following tables furnish the results of an experiment conducted to determine the effect of high temperature on survival of *Penaeus merguiensis* and *P.monodon* of 8.5 cm in length (Piyakarnchana *et al.*, 1975).

Table 1: Temperature Tolerance of Penaeid Shrimps

Penaeus monodon							
Temperature (°C)	26.5	30	35	37.5	40		
Per cent survival (per cent)	100	100	100	60	0		
Penaeus merguiensis							
Temperature °C	30	344	36	38	40	42	42.5
Per cent normal shrimp	100	100	50	50	0	0	0
Per cent shrimp immobile	0	0	50	50	25	25	0
Per cent dead shrimp	0	0	0	0	75	75	100

The best way to ensure that the temperature of shrimp pond water does not become too hot is to provide adequate depth of water. This can be achieved by deepening the total pond area, or by excavating deep channels within the pond for the shrimp to seek shelter in.

The effect of wind action on water movement and mixing is not as great in deep ponds as it is in shallow ponds. Consequently stratification of water layers as in the case of heavy rains may be dangerous for growing shrimps. Hence, ponds should be designed to hold relatively deep water levels during hot dry season. Precautions also need to be taken to ensure adequate mixing of the water.

Shading portions of the pond with floating material such as coconut and palmyra leaves have also been found beneficial. Now a days artificial bio-films and strips are also used for this purpose.

(ii) Salinity

Young shrimps can tolerate fluctuations of salinity. In most species salinity has little effect on either survival or growth of post larvae, except at extremes. The ability to withstand extremely low salinities varies from species to species. The period of acclimation is important in determining the lowest salinity at which a shrimp can survive. Changes in salinity should be as gradual as possible because abrupt exposure to very low salinities may lead to death. Of the important species cultured, it is generally considered that *P. monodon* has wider salinity tolerance. Though the optimal growth of *P. merguiensis* was obtained at 27 ppt, *P. semisulcatus* seems to require a relatively high saline water. All species of penaeid shrimps however require almost marine seawater for maturation and spawning.

To protect the shrimp ponds against abrupt changes in salinity:

(a) There must be a provision to exchange pond water rapidly, and whenever it is required.

(b) Sluice gates must be designed so as to permit rapid draining of surface water during and after heavy rains.

(c) Sluice gates should be designed to permit the inflowing water for replenishment from the bottom at times when the surface water is of low salinity in the adjacent natural waters.

(d) Pond water should be at least 50 cm deep for temperature control. This also aids in control of salinity as the greater water volume provides more protection against dilution. For example, if a pond of 10 cm deep receives 10 cm of rainfall, salinity will drop by 50 percent. If the water column of a shrimp pond is maintained at a depth of 50 cm, 10 cm of rain will reduce the salinity by only 17 percent.

(e) Diversion canals should be provided to divert rain water runoff from adjacent land away from the pond to prevent destruction of dikes and flooding of the pond.

(f) To prevent high salinity resulting from evaporation, windbreaks such as trees or high dikes may be useful. Trees with more or less evergreen leaves should be grown because if a lot of leaves fall into the pond they may cause problems when they decompose.

(iii) Oxygen

Maintenance of adequate levels of dissolved oxygen in the pond water is very important for the growth of shrimps. In culture system, many workers have suggested that the minimal level of oxygen needed for optimal shrimp growth is above 2 ppm, but there are inadequate data to support this conclusion as the DO level is influenced by several factors in an aquatic environment. It has been reported that for *P. japonicus,* stress is signaled at 1.4 ppm when burrowing occurs. It was observed that in *P. schmitti* the majority of shrimp began swimming at the water surface when the level of dissolved oxygen was reduced to 1.2 ppm. When the immobile shrimp were placed in well-aerated tanks, about 50 per cent recovered. Considering the above, perhaps a dissolved a dissolved oxygen level of 1.2 ppm should be considered as a base at which shrimp start to die with even a short exposure.

Regarding the long-term effects of sub-lethal dissolved oxygen levels, it is reported that white areas of degenerated tissue in the tail muscles of shrimps could be associated with low levels of dissolved oxygen and high temperature. Shrimp with this condition normally will be dieing. When the affected shrimps were placed in a well-aerated tanks, the white areas however disappeared within 24 hours and the shrimp became active. The same condition was also observed in *P. monodon* culture ponds.

Aquaculturists feel that when dissolved oxygen level reaches 3 ppm or below in shrimp ponds, remedial measures are essential. The growth of shrimps should be the best at dissolved oxygen levels above 3 ppm and that mortalities will occur after short-term exposure at dissolved oxygen levels below 1.2 ppm. However, this may not be always hold true in a pond where several other factors interact. Mortality can be reduced in shrimps suffering from a lack of dissolved oxygen if the oxygen level is raised quickly.

The concentration of dissolved oxygen in ponds is normally expressed in terms of its percent solubility. It is established that water with a high temperature and salinity holds less dissolved oxygen than the water with low temperature and salinity. Consequently a deeper pond would be beneficial in maintaining reduced temperature and for providing increased oxygen solubility which with proper management could result in increased levels of dissolved oxygen.

Growing shrimps in culture system are not probably the main consumers of oxygen especially at a low level of dissolved oxygen. Shigueno (1975) reported the estimated percentages of oxygen consumed in one night in a shrimp pond with a high organic load was as follows:

Table 2: Consumers of Oxygen in Shrimp Aquaculture Systems

Oxygen Consumers	*Percentage (per cent)*
Shrimps	8.6
Other crustaceans	0.5
Fishes	6.7
Bottom sand	14.8
Water	69.4

Therefore, other components of shrimp culture systems such as algae, bacteria and detritus consume significant quantity of dissolved oxygen. The most effective way to correct low dissolved oxygen levels in such a pond is to reduce the amount of algae, bacteria and detritus in the water. This can be done by draining a portion of the pond water and refilling it with clean water, which is otherwise called as 'exchange'.

Heavy rains can cause stratification of water layers, especially if the pond is deep without much wave action. In such situtation the lighter freshwater floats on top of the more dense salt water. Such stratification can results in oxygen depletion in the lower salt water layer. Provision should therefore be made to promote mixing of water after heavy rains.

Increased water movement provides more aeration which can be used to keep the dissolved oxygen levels from falling to a critical point. The critically low levels of oxygen could be corrected by:

(a) Water exchange, especially letting new water into a pond. Pumping is the only way to do this at the time it is needed.

(b) Installation of aeration equipments

- *Surface aerators*: *e.g.* Paddle wheel aerators, sprinklators, etc.
- *Air injectors*: *e.g.* AIRE–O_2 type aerators.
- Air blower or oil-free compressor with perforated airlines and/air stones provided in the pond/tank
- Electric water mill
- Wind mill based devices

(c) Orientation of the long axis of the pond with the prevailing wind during the construction stage. Caution must be exercised as in areas with strong winds, wave action might cause excessive dike erosion, especially in large, deep ponds, and it might be necessary to provide wind wave breaks near the dike. In such areas it may be more advantageous to orient the short axis of the pond with the prevailing wind and rely on other means of providing aeration.

(d) Construction of large ponds which allow a greater sweep of wind across the pond.

(e) Lowering the water depth to utilize the effect of wind action (care must be taken that sufficient depth is maintained to prevent high water temperature).

(f) Not constructing the dikes excessively high so that they block the wind.

(g) Not planting trees on the dikes.

All the above factors can have an effect on some other aspects of pond management and each factor must be evaluated to assess its effects on the overall scheme in each locality.

(iv) pH

The pH plays a very important role in shrimp aquaculture management and a low water pH can affect the shrimp directly. Even though *P. monodon* grows without mortalities in waters with a pH of 6.0 in the presence of inorganic carbon, growth would be reduced to 60 percent. However, a drop in pH that is associated with a loss or rapid reduction of inorganic carbon, may be lethal. In water with a pH of 6.4, and less than 10 to 12 mg/l of inorganic carbon, penaeid shrimps exhibit significant reduction in growth and lower survival. When pH fell below 5.0, heavy mortalities occurred (Wickens, 1976). A fall in pH may have indirect effects also, for instance, resistance of the shrimp to pathogens might be reduced.

One of the most important causes of low water pH is acid soil. Acid sulfate soils are commonly overlaid by good soil which lies above the mean high water level. If the land is excavated to make the pond bottom at a level where the pond can be filled and drained using tidal fluctuation, acid sulfate conditions develop when the subsoil is exposed. This will result in low pH of the pond water unless the soil is improved. Considering the cost an difficulty required to improve an acid sulfate subsoil, it is suggested that in areas where there is a non-acid topsoil, it may be more economically favourable to use bar ditch type construction and fill the ponds by pumping.

During the formation of pond, the surface layer of good soil can be set aside and replaced as a surface layer on the pond bottom and dikes. If the amount of good soil is limited, it should be used to surface the dikes into the pond water. This is much more critical in small ponds than in large ponds. Pond bottoms can be leached or limed to reduce or eliminate the acid condition. In areas where there is not enough good topsoil to surface dikes, the dikes can be made with a berm, and a ditch can be cut in the berm to catch acid water runoff and prevent it from contaminating the pond water. High pH may also lead to ammonia toxicity as it increases the ratio of toxic unionized ammonia in solution to the total ammonia present.

(v) Nitrogen Compounds

Nitrate

The growth of *P. monodon* was not found to be affected by a concentration of 200 mg NO_3–N/l even after three to five weeks of exposure.

Nitrite

In *F. indicus,* growth was reduced by nearly 50 percent over a period of 35 days when nitrite concentration was 6.4 mg NO_2–N/l.

Ammonia

Chronic toxicity tests for ammonia were conducted with five species of penaeid shrimps *viz. P. japonicus, P. occidentalis, P. schmitti, P. semisulcatus* and *P. setiferus.*

The tests showed that a mean concentration of 0.45 mg NH_3–N/l reduced growth by 50 percent of that of controls. It was estimated that a "maximum acceptable level" at which growth would be reduced by only 1 to 2 percent is 0.10 mg NH_3–N/l.

As it is more convenient to measure ammonia in terms of total ammonia nitrogen (not free NH_3 or unionized ammonia), Wickins compiled the following details (Table 3) to give values of total ammonia nitrogen which correspond to the value 0.10 mg unionized ammonia (NH_3-N) per litre at selected temperatures, salinity and pH.

The concentration of total ammonia nitrogen (in mg/l) that corresponds to a calculated level of 0.1 mg/l unionized ammonia nitrogen in water at a constant pressure of 1 atmosphere at different values of temperature, salinity and pH (Wickins, 1976) are as follows:

From the table it can be seen that pH has a major effect, with the percentage of toxic unionized ammonia being much greater at high pH than at low pH. In water with a temperature of 28°C, salinity of 24 ppt and pH of 6.8, the critical level of 0.1 mg/l unionized ammonia occurs when the total ammonia level is 26.1 mg/l. In water with a temperature of 28°C, salinity of 24 ppt and pH of 8.4, a level of 0.1 mg/l unionized ammonia occurs when the total ammonia level is only 0.8 mg/l.

The normal pH of brackish water is 8.0 to 8.3 and in ponds with a good growth of phytoplankton, pH values of 9 and above are common in the late afternoon. Most of the ammonia in a pond is formed due to the waste accumulation of the organisms originated or living in the pond. The higher the density of both the species being

Table 3: Total Ammonia Nitrogen Levels Vs Other Factors

Salinity	*0 per cent*		*24 per cent*		*27 per cent*		*30 per cent*		*33 per cent*	
Temperature	*20°C*	*28°C*	*20°C*	*28°C*	*20°C*	*28°C*	*20°C*	*28°C*	*20°C*	*28°C*
pH	*(Concentration of total ammonia nitrogen (mg/l))*									
6.8	40.4	22.3	47.4	26.1	48.4	27.7	49.4	27.2	50.4	27.8
7.0	25.5	14.1	29.9	16.5	30.6	16.9	31.2	17.2	31.8	17.6
7.2	16.2	18.9	18.9	10.5	19.3	10.7	19.7	10.9	20.1	11.1
7.4	10.2	2.7	12.0	6.6	12.2	6.8	12.5	6.9	12.7	7.1
7.6	6.5	3.6	7.6	4.2	7.8	4.3	7.9	4.4	8.1	4.0
7.8	4.1	2.3	4.8	2.7	4.9	2.8	5.0	2.8	5.1	2.9
8.0	2.6	1.5	3.1	1.7	3.2	1.8	3.2	1.8	3.3	1.9
8.2	1.7	1.0	2.0	1.1	2.0	1.2	2.1	1.2	2.1	1.2
8.4	1.1	0.7	1.3	0.8	1.3	0.8	1.3	0.8	1.4	0.8

cultured and the organisms cultured for food, the greater the production of ammonia. Ammonia will eventually be converted to nitrate, but there is a danger that ammonia production will exceed the capacity of the pond to convert the ammonia rapidly enough to prevent it from exceeding toxic levels. Some species of planktonic algae such as *Chlorella sp. Chaetoceros sp.* etc. can utilize ammonia and nitrate directly. If these organisms are present in optimum levels the danger of ammonia build-up will be reduced. However, it is very difficult to control the species of algae growing in a pond.

An additional factor is that when dissolved oxygen levels are low, nitrates are reduced to ammonia, thus increasing the level of ammonia in the water. A decrease in dissolved oxygen also increases the toxicity of unionized ammonia. Conversely, an increased level of dissolved oxygen reduces toxicity.

(vi) Hydrogen Sulphide

Hydrogen sulphide (H_2S) in a pond is produced by the chemical reduction of organic matter which accumulates on, or in, the pond bottom. The bottom soil turns black and sometimes a rotten smell is discharged. As shrimps live primarily on, or in, the bottom, a build-up of H_2S in the bottom soil, or in water near the bottom, is inevitable. It has been determined experimentally that shrimp (*P.japonicus)* lost equilibrium when exposed to a level of 0.1 to 2.0 ppm hydrogen sulphide (H_2S) in water. Shrimp died instantly at a concentration of 4 ppm.

Studies in one pond showed that the concentration of sulphide-sulphur (mostly H_2S) in interstitial water, 2 cm deep in the pond bottom reached as high as 10 ppm. In the pond water it exceeded 0.09 ppm, varying from 0.037 to 0.093 ppm. In other condition shrimps normally face mortality. Shrimp buried in the bottom draw in water from above the pond bottom, so it is not likely that the level of H_2S in the water was lethal. Dissolved oxygen level in the water did not fall below 2.7 pm, again beyond the lethal limit.

To determine the longer term effects of H_2S on shrimps, soil on the bottom of cement tanks can be treated with an application of iron oxide (70 per cent ferrous oxide (Fe)) at the rate of 1 kg per m^2. The shrimp in the tank with the treated bottom grow significantly better than those in the untreated tank.

However, it would not be practical to treat bottoms of large ponds with ferrous oxide, but frequent changes of water would prevent, the build-up of H_2S in the pond water. If ponds are constructed with peripheral canals, treatment of only the canals with iron oxide might be practical, as most of the organic debris is deposited in these canals and H_2S production should be restricted in these canals only.

Chapter 4
Stocking Management

A. Pond Preparation

(i) Drying the Pond Bottom

Drying the pond bottom periodically in shrimp farms help to mineralize the organic materials accumulated in the soil. It also reduces the production of H_2S and other harmful substances that may be produced during anaerobic reduction of the organic material when the pond is with water. If a pond is completely dried, all unwanted predators and competitors are destroyed and there is no need to treat with chemicals to get rid of them. Drying the soil is especially useful in ponds where lab-lab, heterotrophic bacteria and other plankton are grown for food and bioremediation. The firm soil provides a good surface for algal attachment.

A word of caution is needed concerning the drying of pond bottom in areas of high acidity levels. In pyrite soil, during the drying process, parameters are optimized and when the pond is filled with water acid is formed and pH of the water is lowered. This type of pond should be flushed thoroughly after drying.

It is suggested that for lab-lab production in ponds with hard bottom, the soil should be tilled after drying. Tilling of soil helps largely in the mineralization of organic matter. However oxidization of acid sulfate soils may be harmful, particularly where fresh soil is exposed to the air. Hence tilling is not recommended for acid sulfate soils. Similar tilling of non-acid pond bottom soil on a regular basis is also not recommended. Tilling should be done only when the bottom is hard and earlier yield was low.

Excessive drying seems to be harmful, and over-drying which results in crumbling and reduction of the thin surface crust to powder is to be avoided.

Following are some procedures recommended for drying of pond bottom:

(a) Drying for seven days

(b) Drying to a point where bottom soil surface will not sink to 1 cm

(c) Drying until the top 1 cm is dried

(d) Drying until the soil cracks 1 to 2 cm deep

In ponds with internal canals, the accumulated mud and organic debris must be removed periodically. Cleaning and deepening of the internal canals should not be done before harvesting shrimps in the pond. The large amount of H_2S released by digging the pond may be harmful to shrimps. Hence digging of ponds should be done while it is dried. The excavated sediment is usually heaped on the dike carefully. It is also advisable to cover the formed dikes with grass turf so that the grass would arrest movement of materials into the pond with the first heavy rain.

(ii) Improving or Controlling Soil Acidity

In ponds where water is not changed frequently, soil pH should be at least 6.5 for proper management. Ponds with a soil pH lower than 6.5 can be managed only as long as frequent water changes are made. A change of water is required at least every three days. One way of improving ponds with acid sulfate soil is to repeatedly dry the pond and then flush it by repeatedly filling and draining. Acids formed by pyrite oxidation will gradually be removed by this process. After a pond is dug in an acid soil area, it should be flushed well until no, or little, red coloured scum from oxidized iron is observed. Lime should be added only after the pond is flushed.

(iii) Liming

Lime can be used to control soil and water acidity. If it is not possible to perform the soil test to determine the correct amount of lime to be added, the soil pH can be tested as follows. Due to the high cost of treatment, applying agricultural lime may not be advantageous when the soil pH is very low, ie. less than 2.5. For soil with a pH of 5, treatment with 3 tons per ha of agricultural lime has been found to be effective. When lime tailings are used (from hydrate of lime processing), only one-half of the recommended dose of agricultural lime is used. While applying the lime it should be worked into the soil. This can be done with a hand pulled harrow. Alternatively pond bottom may be treated by broadcasting 1.5 tons of agricultural lime per hectare. The bottom is then leveled and another 1.5 tons is worked into the soil.

Application of lime of calcium carbonate (calcite) may not be suitable particularly when pond water attains seawater pH. Though agricultural lime may raise the soil pH, it will not be effective in maintaining suitable pH in the pond water. As natural carbonates (*e.g.* dolomite, mollusc shells or coral) which contain about 4 percent magnesium are more soluble at seawater pH and they could aid in maintaining optimum alkalinity and pH levels in the pond water (King, 1973).

If the dikes are made up of acid sulfate soils, proper water management should be undertaken in order to reduce the problems associated with them. By maintaining equal water levels in ponds and keeping this level higher than water levels in the canal system, the transfer of acids and active aluminium and iron into ponds by seepage through dikes can be limited. Proper control of the water table in drained pond soils can be used to limit the proper depth of soil drying, thereby limiting pyrite oxidation and acid formation.

Controlling erosion to prevent acid runoff into the pond is especially important when the dike soil is acid sulfate in nature or when material from internal canals is thrown on the dikes during cleaning and deepening. Acid tolerant African Star Grass (*Cynadon plectostachus*) provides good vegetative cover for dikes. Other *Cynadon* species are also worth trying. The following procedure is recommended to establish grass on acid soil. Planting should begin at the start of the rainy season. First, the soil should be tilled to a depth of 5 cm. Then suitable quality agricultural lime are added according to soil acidity. Five tons per ha of chicken manure and 35 kg per ha of 14-14-14 (NPK) are added subsequently. The prepared area is covered with a 5 cm thickness of rice straw. Cuttings are then planted at 30 cm intervals.

(iv) Poisoning Predators and Pests

Before shrimp PL/juveniles are stocked, eggs and larvae of competitors such as fishes, crabs etc. should be killed by poisoning. It is paramount importance because most of these organisms have also been proved to be the carriers of white spot syndrome virus.

(v) Biofeed Production in Ponds

(a) Lab-lab

"Lab-lab" is characterized mostly by benthic blue-green algae and diatoms, but many other forms of plants and animals are also associated with it. For good growth, "lab-lab" requires low water levels from 5 to 40 cm. Best growth is reported to be at salinities of 25 ppt or higher.

The requirement of "lab-lab" for high salinity is not compatible with optimum growing conditions for *P. monodon* which is reported to grow best at slightly lower salinities (10–25 ppt). Therefore it is well suited for *F. indicus* or *P. merguiensis*.

For the better utilization of "lab-lab" in shrimp culture, the following aspects need to be kept considered.

1. "Lab-lab" can be used for shrimp culture during the first two months of culture or up to a point when the shrimps grow to a size of about 3–4g. It can be assumed that "lab-lab" might be a suitable food to grow in a nursery pond.

2. Ponds can be constructed with a large number of interior canals atleast 1.5 m deep to provide shrimp with shelter against high temperature during the day. As shrimps largely feed during night hours, the shallow portions with "lab-lab" may serve as feeding platforms on which the shrimp could graze.

Method of Growing and Managing 'Lab-lab'

Soil with a high clay content support the best growth of "lab-lab". The relationship between soil texture and algal growth can be seen in Table 1.

Table 1: Soil Quality Vs. Benthic Fauna

Sl.No.	*Per-cent Sand*	*Per-cent Silt*	*Per-cent Clay*	*Soil Texture*			*Growth of Benthic Algae*
1.	28	22	50	Clay	–	–	Very abundant
2.	15	44	42	Silty	Clay	Loam	Abundant
3	63	14	23	Sandy	Clay	Loam	Few
4.	79	10	11	Sandy	Loam	–	Very few

Preparation of the pond soil is very important in growing "lab-lab". To assure a uniform growth of algae, the pond bottom should be leveled so that there are no high points or depressions. The pond bottom should not be hard but must be firm enough to serve as a holdfast for the algae. Compacting the pond bottom is normally done by drying. The bottom should not be bone dry. It is better to dry it just until a man can walk on it without sinking in. It usually takes 7 to 10 days of drying to reach this point.

Growth of "lab-lab" is also directly related to the amount of organic matter present in the soil. The following relationship of organic matter to the growth of algae could be seen normally.

Table 2: Organic Matter Vs. Algal Growth

Organic Matter (per cent)	*Growth of Algae*
Above 16	Very abundant
9–15	Abundant
6–8	Few
6	Very few

To increase the amount of organic matter in the soil, fertilizer, chicken manure or other manures are applied to the dry pond bottom at the rate of 350 kg/ha. The chicken manure should be dried and not treated with insecticide. Instead of organic manure, inorganic fertilizer can also be used at the rate of 50–100 kg of 18-40-0 (N-P-K) or 100–150 kg of 16-20-0 per hactare.

Immediately after fertilization, 3 to 5 cm of water is let into the pond. After one week, the same amount of fertilizer is applied and the water level is raised to 10 to 15 cm. The fertilization is repeated after the second week and the water level is raised to 20 to 25 cm. Additional water may be added to make up for compensating loss due to evaporation. It is also recommended for refertilization once in seven days during the culture period.

(b) Phytoplankton

The phytoplankton is composed of small plant matters which float in the water. While phytoplankton does not serve as a direct food for shrimps, its production level indicates the functionality status of shrimps. Further, phytoplankton growth leads to zooplankton production in ponds and better phytoplankton and zooplankton production would be responsible for the development of debris which serves as food for growing shrimps.

Phytoplankton production is better in ponds with a water level of 70 cm or more, depending on management practices adopted which include the type of fertilizer used, salinity maintained, etc. Optimal phytoplankton growth would give yellow–green or yellow–brown colour to the pond water. Heavy mortality of shrimp has occurred in ponds when pond water showed a bright green or reddish colour. This may possibly be due to the obnoxious phytoplankton blooms particularly of toxic dinoflagellates and other forms of plankton.

Method of Growing and Managing Phytoplankton

In shrimp culture the benefits of fertilization are indirect to the cultivable shrimps. Fertilization causes a good growth of phytoplankton and the resulting microorganisms feed on the phytoplankton and the shrimps feed on such microorganisms. It has been observed that growth of shrimp is better in ponds in which diatoms are abundant and poor growth has been observed in ponds in which the predominant algae were phytoflagellates. These two types of phytoplankton have different nutrient requirements. In laboratory and tank culture, nitrogen (N) to phosphorus (P) ratios of 20 or 30 to 1 have been found suitable for diatoms and ratios close to 1:1 for phytoflagellates. The same nutrient requirements should also hold true for algae growing in ponds. The following table provides the levels of nitrogen and phosphorous required in shrimp ponds for optimal growth of phytoplankton.

Table 3: Level of Application of Nitrogen and Phosphorous

Nitrogen (ppm)	*Phosphorus (ppm)*
1.4	0. 15
1.3	0. 14
1.1	0. 12
0.95	0. 11
0.8	0. 09
0.7	0. 08
0.6	0. 07
0.4	0. 05
0.3	0. 03

One of the most important factors to be considered in a programme of fertilization is that both nitrogen and phosphorus do not remain in solution for very long duration after they are added to the pond water. They become incorporated in living organisms or into the bottom soil. This is especially true for nitrogen as larger amounts are added. Following the application of ammonium-bearing fertilizer, most of the added nitrogen would be absorbed by colloids in the bottom soil within a few days and remained strongly bound there. The amount of nitrogen absorbed in the bottom soils will be quite small when a nitrate fertilizer is added, and a higher level of available nitrogen is maintained in the water. In selecting the form of nitrogenous fertilizer, (ammonium or nitrate), to be used in salt water pond, consideration should be given to the type of organisms to be cultured as food. If phytoplankton is to be grown, nitrate fertilizers would be better. If bottom growing organisms such as blue-green algae are to be cultured, ammonium-based fertilizers would be recommended.

As a great portion of the nutrients added to a pond is known to be absorbed in the soil after a short time, frequent applications of small amounts of fertilizer would give the best results. Application of fertilizers every 7 to 10 days is normally recommended. As the nutrient level of seawater varies from place to place and season to season, a programme of fertilization that works successfully in one location might not be working well in another area. Hence, it is necessary to apply fertilizers at different rates/levels at different times of the year in shrimp ponds.

The best way to develop a suitable method for fertilizing pond water is to apply a moderate amount of fertilizer and observe its effect on the phytoplankton growth. Subsequently depending on the performance, the rate of application is to be adopted or determined. To assess the density of plankton developing in the pond, a ' secchi disc ' can be used. When the secchi disc reading is about 30 cm, the phytoplankton density would be good. If the secchi disc disappears from sight at less than 20 cm, the phytoplankton may be dense and the change of pond water is to be undertaken. Further, the subsequent application of fertilizers should be reduced. If the secchi disc disappears from sight beyond 35 cm, phytoplankton growth is not enough and more fertilizer should be added during the next application. Eventually a farmer will learn how much fertilizer is required to maintain a good growth of phytoplankton in the pond.

(c) Benthic Animals

Heavy application of organic fertilizers encourages the growth of chironomid larvae which help in the good growth of shrimp. A juvenile shrimp weighing 0.06 g is known to consume 23 chironomid larvae in 24 hours. Dense populations of chironomids are often associated with low levels of dissolved oxygen and care must be taken in encouraging their growth in the pond. Further, heavy population of chironomids may also down lab-lab.

Estimation of the Desired Fertilizer Application Rate

It is assumed that a level of 0.95 ppm nitrogen and 0.11 ppm phosphorus should be suitable as a starting dose. The following method can be used to estimate the amount of nutrient required to achieve these levels. First, estimate the volume of water in the pond. For example, a one hectare pond has a surface area of 10,000 m^2. If it has an average water depth of 60 cm, the volume of water in the pond would be 10,000 $m^2 \times 0.6$ m = 6,000 m^3. One ppm is equal to 1 gram per m^3 of water. So to find out the amount of nitrogen which should be added to the one hectare pond to get a level of 0.95 ppm, the volume of water is multiplied by 0.95 g, thus:

$$6000 \times 0.95 \text{ g} = 5700 \text{ g or } 5.7 \text{ kg N}$$

The quantity of phosphorus to be added is found in the same manner.

$$6000 \times 0.11 \text{ g} = 660 \text{ g or } 0.7 \text{ kg P}$$

Once the amount of nutrient required is determined, the amount of fertilizer which contains the desired amount of nutrient can be determined, as follows:

$$\text{Fertilizer to be added (kg)} = \frac{\text{Amount of nutrient desired}}{\text{Percent nutrient in the fertilizer}}$$

If the pond is to be fertilized with ammonium sulfate which contains 21 percent nitrogen, then the quantity of ammonium sulfate required is as follows:

$$\frac{5.7}{0.21} = 27.\ 1\text{kg.}$$

Triple superphosphate contains 39 percent phosphorus, so following the same procedure, the amount of triple phosphate required would be:

$$\frac{0.7}{0.39} = 1.9 \text{ kg.}$$

The percent nitrogen (N) in some common fertilizers are:

Urea–CO (NH_2): 46.6 per cent

Ammonium sulfate–$(NH_4)_2$ SO_4: 21 per cent

Ammonium chloride–NH_4 Cl: 25 per cent

Ammonium nitrate–NH_4 NO_3: 37 per cent

Calcium nitrate–Ca $(NO_3)_2$: 17 per cent

The percentage of phosphorus (P) in superphosphate fertilizers are:

Double super phosphate–Ca ($H_2 PO_4$): 26 per cent

Triple super phosphate–$P_2 O_5$: 39 per cent

Many fertilizers contain more than one primary nutrient. In these, the primary nutrients are designated by a numbering system indicating percentage in each nutrient. The numbering system is always listed in the following order: N (nitrogen), P (available phosphoric acid $P_2 O_5$), and K (potash $K_2 O$). K is usually present in sufficient quantity in brackish water and hence it is not normally added.

By referring to the numbers printed on a fertilizer bag, the types of nutrients and their levels could be known. For example:

12–24–12 contains 12 per cent N, 24 per cent available $P_2 O_5$ and 12 K.

16–20–0 contains 12 per cent N, 20 per cent available $P_2 O_5$ and no K.

45–0–0 contains 45 per cent N, no available $P_2 O_5$ or K.

0–0–60 contains no nitrogen or available $P_2 O_5$, but has 60 per cent K.

Since these numbers are percentages, a 50 kg of 12–24–12 would contain 6. 0 kg N, 12.0 kg available $P_2 O_5$ and 6.0 kg of $K_2 O$. As $P_2 O_5$ contains only 44 percent P, the weight of P is 4.7 kg.

It is not compulsory to use only inorganic fertilizers organic fertilizers can also be used. The percentage of N and P in few organic fertilizers are given in Table 4. Additional quantity of N or P may be required to obtain maximum benefits from the organic fertilizer.

Table 4: Algal Production Vs. Different Fertilizer Combinations

Fertilizers	*Lab-lab*	*Plankton*
Urea	623. 0	475. 5
Chicken manure + Urea *	514. 0	826. 7
Chicken manure + ammonia phosphate **	424. 0	721. 3
Chicken manure + phosphate ***	878. 3	341. 7
Ammonium phosphate	339. 5	451. 7
Chicken manure	468. 0	312. 3
No fertilizer	346. 5	190. 8
Phosphate	382. 5	172. 5
Means	497. 0	437. 3

*: 46–0–0; **: 16–20–0; ***: 0–20–0.

It can be seen that highest production in those ponds managed for plankton could be obtained when additional N is supplied. Conversely, highest production in the ponds managed for lab-lab could be obtained when additional P is added. Irrespective of the species cultured high plankton production will be associated with more N in relation to P.

Mode of Fertilizer Application

The platform method is an effective way to apply inorganic fertilizers to ponds for producing good growth of phytoplankton as the nutrients from the fertilizer on the platform can be released slowly and distributed through out the pond by water movement. The platform should be positioned so that its top surface is about 15 to 20 cm below the water surface, and located near the end of the pond from which the prevailing wind comes. A single platform is sufficient for pond up to 0.7 ha when plankton is grown.

Suggested platform top sizes for ponds of different sizes are:

Table 5: Pond Size Vs. Platform Size

Pond Area (m^2)	*Platform Top Dimensions (m)*
1000	0. 85 × 0. 85
2000	1. 25 × 1. 25
3000	1. 50 × 1. 50
4000	1. 70 × 1. 70
5000	1. 90 × 1. 90
6000	2. 10 × 2. 10
7000	2. 25 × 2. 25

The fertilizer to be applied is simply piled onto the platform and left alone.

Recommendations for Managing Natural Food Organisms in Shrimp Ponds

(a) Start with phytoplankton or "lab-lab" for no more than the first two months after post larvae are stocked.

(b) The shrimps should be held afterwards in ponds where production of phytoplankton or "lumut" is maintained depending on the salinity level.

(c) This can be accomplished by growing the shrimps in a nursery pond for the first two months and then transferring them to another pond. The second method is to keep the water level in a pond low for the initial two months provided temperature do not raise drastically and then raising the water level sufficiently to encourage the growth of other types of algae.

B. Identifying Shrimp Larvae

Characteristics which are used to distinguish adult penaeids can also be used to identify their post larvae. In penaeid shrimps, the first three pairs of walking legs have chelae (pinchers) and the first abdominal segment overlaps the second. However these characteristics can be identified using a microscope in post-larvae. Probably, the aquatic animal most commonly mistaken for post-larval shrimp is *Acetes spp.* (Sergestidae). They can be easily distinguished by their long, bright orange antennae which have a prominent sharp bend in them. Post-larvae of shrimps have short, colourless antennae. In addition, these larvae do not have statocysts (in live animals these appear like small bright spots to the naked eye) on the tail (Mysidacea), and their eyes do not extend laterally at a 90° angle (Sergestidae). The uropods of Mysidacea are not spread fan-like as in shrimp, but are parallel, directed posterior.

'Penaeus' post-larvae are long and thin, while that of 'Metapenaeus' are relatively short and stout. Metapenaeus are generally coloured a mottled gray or brown. Post-larvae within the genus Penaeus are generally almost colourless, however, *P. monodon* and *P. semisulcatus* are coloured rust-brown. When the pigment chromatophores expand, a prominent bluish or reddish-brown streak appears on the ventral side of the body in *P. monodon* postlarvae. In *P. semisulcatus*, the chromatophores are prominent only on the sixth (last) segment. *P. monodon* is larger than *P. semisulcatus* and has a distinctive habit of swimming with its head lower than its tail and the body is kept at about 45°.

The post-larvae of *P. monodon* and *P. semisulcatus* can be separated by the following characteristics.

Table 6: Characteristic Differences Between *P. monodon* and *P. semisulcatus*

P. monodon	*P. semisulcatus*
• Brown pigment distributed evenly but more intensively on posterior part of telsion.	• Brown pigment distributed only on the anterior base and about one-third posterior tip of telsion.
• Endopod of uroped may be totally or partly pigmented.	• Only posterior part of endoped of uropods is pigmented.
• Exopod of uropod is mostly unpigmented; in some cases with a small blotch on the inner latero-posterior margin.	• Exopod of uropod mostly with small blotch on the inner later oposterior margin as continuation of pigmentation of inner plate.

Provisional Key for the Identification of Post-larvae of *Penaeus monodon, F. indicus* and *P. semisulcatus* (Subrahmanyan and Rao, 1970)

(a) Five to seven reddish-brown (or yellowish) chromatophores on the ventral side of the sixth abdominal somite (segment). One reddish-brown chromatophore at the anterior end of the six abdominal segment laterally.

.... *Fenneropenaeus indicus*

More than eight chromatophores on the ventral side of the sixth abdominal somite. Present or absent of one reddish-brown chromatophore at the anterior end of the sixth abdominal segment.

(b) Eight to 11 reddish-brown (sometimes bluish) chromatophores on the ventral side of the sixth abdominal somite. One reddish-brown chromatophore at the anterior end of the sixth abdominal segment laterally. One or two reddish-brown chromatophores are also present on the dorsal side of each abdominal segment.

.... *Penaeus semisulcatus*

14 to 19 reddish-brown (sometimes bluish) chromatophores on the ventral side of the sixth abdominal segment. No lateral chromatophore on the sixth abdominal segment anteriorly. The ventral chromatophores on the sixth abdominal segment appears as a bluish or reddish-brown streak in expanded condition .... *Penaeus monodon*

Transportation of Shrimp Seeds

The most common method of transporting shrimp seeds is in plastic bags filled with oxygen. The following points should be considered while using plastic bags for shipment.

(a) Thin polyethylene bags are permeable to oxygen. So when post-larvae are to be shipped for longer duration, better results can be obtained by placing the bags in an impermeable container and sealing the container tightly. If polyethylene bags are used, shrimps should not be held in the bag for more than six hours without a change of water and oxygen filling.

(b) Post-larvae should be held without food for several hours before packing. Then they should be screened thoroughly and placed in clean water to eliminate as much trash as possible. Decomposition of the trash depletes oxygen and the trash itself serves as nutrients for harmful bacteria. Small amounts of activated charcoal may be added to the bags to absorb harmful waste materials produced by the shrimp.

(c) As the polythene bags are prone to be punctured double bags should be used as a safety measure.

(d) Containers holding seeds should not be exposed to the sun, but kept in the shade. It is better to make shipments at night or cool hours.

The most commonly used polyethylene bag is of 50 × 75 cm size. Five to six liters of water is filled up in the bag. Then the shrimp seeds are added, as follows:

Total length 10 mm (shrimp): 15,000

Total length 10–20 mm (shrimp): 5,000

Total length 20–25 mm (shrimp): 3,000

The bag is then closed so there is no air inside the bag. Sufficient oxygen is added to fill the bag which is sealed with elastic bands or string. The bag is then placed in a container which is impermeable to oxygen. Styrofoam is preferable if it is available. Two fist size pieces of ice in a plastic bag are placed in the container alongside the bag holding the post-larvae. The ice bag is to cool the water and reduce metabolism of the shrimp. The container is then sealed with tape. By this method shrimps can be transported at a survival rate of more than 90 per cent.

Holding and Growing Small Post-Larvae to a Larger Size for Stocking

It is better to hold and grow shrimp seeds to a size of 2.5 to 3.0 cm instead of stocking small post larvae directly into the rearing ponds. This can be done in well-prepared nursery ponds or tanks. Hapa nets have been found to give poor survival. It is better to stock directly into the pond than using hapa nets for nursery rearing.

One of the problems with stocking directly into a rearing pond is that it is almost impossible to assess their survival at that stage. A farmer for a catchable size has to wait until the shrimps grow before he knows whether they are alive or otherwise. A good practice is to keep a few post-larvae in a container with pond water and watch them for several days. If these seeds die, the farmer should check his pond extra carefully and arrange for restocking, if necessary. This practice is particularly recommended for seeds obtained from hatcheries, as the chance for disease incidence is greater in hatchery-produced seeds.

Nursery Ponds

When nursery ponds are used, the shrimps should be acclimatized gradually to pond conditions to prevent death or damage from the shock of rapid change in temperature or salinity.

Some farmers keep the seeds in plastic shipping bags and float the bag in the pond water for a short time to acclimatize them. In cases where temperatures are nearly the same and the shrimps are healthy, the floating technique will reduce losses. However, if they have been subjected to low oxygen levels and high levels of carbon dioxide and ammonia, especially during longer trips, it would be more harmful to keep them in the bags exposed to unfavourable conditions.

It is better to use a tank with aeration for acclimatization. Water in the tank should be adjusted to near the temperature and salinity of the water in which the shrimps were transported. After the post-larvae are added to the tank, the water in the tank is gradually adjusted to pond salinity and temperature. The period of adjustment depends on how much the temperature and salinity need to be changed, but usually a period of 12 hours would be sufficient. The fry should not be released into a nursery pond during hot hours of the day. Evening hours would be best suited for this job. The PL can be stocked at fairly high densities in nursery ponds, up to 25 per square metre, or 2,50,000 per hectare. The young shrimps should be kept in a nursery pond for two to three weeks, or until they reach an average size of 25 mm. Further, raceway nurseries are now introduced in countries like USA, Malaysia and now in India, and seeds can be stocked at a very high densities (1000–3000/m^3). Discussed in detail in another chapter.

Transferring Fry from Nursery to Growout Ponds

It is difficult to harvest shrimps or transfer from one pond to another. Transfer should be done by making them move with water flow. This is not always easy to accomplish, especially with *P. monodon*. The following methods are suggested for inducing shrimps to move with the flow of water during transfer from one pond to another:

(a) Transfer at night, and use a light to attract the shrimp to the sluice gate.

(b) Let a little fresh water into the pond preceding the transfer. This makes the shrimps more active. The pond is then drained.

(c) Change pond conditions to make the shrimp become active and ready to move out of the pond. One way of doing this would be to lower the water level so the temperature of the pond is slightly enhanced.

If the growing pond is not adjacent to the nursery pond, the shrimps must be caught and transported. They can be caught in the out flowing water with a minimum of injury by using a suitable net for this purpose, and this net is fastened to a wooden frame which is placed on the sluice gate. As the water is drained from the nursery pond, the young shrimps are caught in the floating catch box. They can be scooped out periodically and transferred to suitable containers for carrying to the grow out pond.

Chapter 5
Shrimp Nutrition and Feed Management

Nutrition and health are two concepts closely linked with one another. Nutritional studies will follow at least these two parameters in order to quantify the requirement of a specific nutrient that is the growth rate and survival. When monitoring survival rate the nutritionist obviously agrees that nutrition does affect health status.

Ideal conditions seldom prevail in outdoor shrimp culture systems as growing animals are often under stress and are confronted with pathogens, and unfavourable environmental conditions. Under these conditions, the requirement for certain essential nutrients will almost certainly increase due to an increased activity of the immune system. Studies related to the effect of micro-nutrients on disease resistance have assessed the possible requirements for these nutrients under problematic situation (Blazer, 1992; Lail and Olivier, 1993).

The economical criteria for selecting specific ingredients in the preparation of feed are important. The addition of higher levels of micro-nutrients or immunostimulants will seriously increase the feed cost ie. upto 200 per cent. The final decision therefore will be in the hands of the farmer and the feed formulator. The feed additives generally improve the final output of the commercial shrimp culture operation. However, it is difficult to conclude as the results are highly variable and do not always allows to draw a clear conclusion.

Feed Formulation, Processing and Quality Control

Manufacturing a feed for shrimp involves several steps and all these steps have a defined effect on the quality and performance of the final product, that is feed. We can classify these factors as follows: (*i*) selection of raw materials, (*ii*) formulation and (*iii*) processing.

When the production process is done according to the basic rules, it will not significantly affect the quality of the diet. On the other hand, selection of raw materials and formulation do have a decision influence on the performance of the diet. Imbalances in formulation or presence of toxins in the ingredients can severely affect shrimp health status and can lead to increased susceptibility to diseases. Processing of feed essentially shapes a particle and gives the ingredient mash the physical presentation that is required for handling and optimal feeding of the shrimp.

Imbalances in Feed Formulation

The effect of feed formulation on the health status of shrimp is rather indirect. Feeding with an inadequately formulated feed to the shrimp will stress its metabolism and will decrease its resistance to diseases. The same would also apply for husbandry, handling, feeding procedures, environmental conditions etc. Shrimp kept in water that is insufficiently oxygenated and containing high metabolite levels will obviously be affected and eventually die due to exposure to unfavourable conditions.

Reviewing all possible imbalances in feed formulation, it would be better to review the nutrition knowledge in its totality. Therefore to formulate a balanced feed, its essential to have a closer look at the effect of ingredient quality or the toxicity of special additives.

Ingredient Selection and Quality Criteria for Feed Formulation

A study with turbot (Obach and Laurencin, 1992) for example, showed that the presence of oxidized lipid (peroxide value: 3000 meq/kg) significantly lowered the chemoluminescent response of head kidney phagocytes and increased mortality rates of the fish when challenged with *Vibrio anguillarum.*

Commercial feeds or feedstuffs having peroxide value higher than 10 meq/kg (acceptable level) are still common. Feed ingredients of marine origin such as fish meal or fish oil, are particularly prone to oxidation due to their high content of highly unsaturated fatty acids. Adequate levels of antioxidants (*e.g.* 200–400 ppm ethoxiquin) should be added in feed preparation. Moreover, if these ingredients are not to be used immediately, their oxidation level should be checked regularly and eventually more antioxidant might be added.

Toxicity of oxidized fish oil can be neutralized by the addition of adequate levels of vitamins E and C. Levels of 100 to 150 ppm are commonly used in commercial fish and shrimp feeds. However, for salmon diets containing high levels of unsaturated fat (upto 35 per cent), levels of 350 ppm of vitamin E or more is used. A level of 40 ppm vitamin E was insufficient to compensate the toxicity of oxidized fish oil in European sea bass (Stephen *et al.*, 1991), while a level of 300 meq/kg) significantly lowered the chemoluminescent response of head kidney phagocytes and increased mortality rates of the fish when challenged with *Vibrio anguilirum.*

Biogenic Amines

Microbial degradation of feed ingredients, especially fish-based, results in the possible production of toxic metabolites (Ruiter, 1995). Biogenic-amines are believed to be problematic substances such as tyramine, histamine, phenylethylamine as they can affect the metabolism of the animal in general and increase blood pressure in particular. Moreover, histamine may bind with lysine and form the toxin gizzerosine during processing under pressure or at high temperatures. Gizzersine is known to provoke gizzard erosion in chickens, but can also cause mortalities in fish and shrimp (Gruz-Suarez *et al.*, 1994). Levels of histamine lower than 500 ppm and levels of gizzersine lower than 1-1.4 ppm are normally recommended in feed formulation.

Mycotoxins

Excess humidity level improves drying of feedstuffs,

especially from vegetable orgin may lead to fungal growth. Some feed ingredients like tapioca, which is mostly sun dried, are often contaminated with aflatoxins. Some of these fungi (*e.g. Aspergillus flavus*) will produce mycotoxins such as aflatoxins, zearaleone, fumonisin, ochratoxin, etc of which aflatoxin B1 is probably the most toxic. Levels of aflatoxin B1 of 50 to 400 ppb are known to affect feed consumption and growth of shrimps besides reduing the resistance to diseases (Hussain *et al.*, 1993; Van Gulick, 1993; Ostrowski-Meissner *et al.*, 1995). *Fusarium moniliforme* (a corn mold) has been responsible for decreased weight gain and reduced resistance to bacterial infection in channel catfish (Lovell *et al.*, 1994). Levels as low as 20 ppm of fumonisin can be toxic to shrimps.

Antinutritional Factors (ANF) from Vegetable Ingredients

Antinutritional factors (ANF) in vegetable ingredients are rather complex as there is a vast array of toxic substances (Tacon, 1985, 1987). Insoluble fibres, soluble fibres, enzyme inhibitors, saponins, lectins, tennins, phytic acid and gossypol are the most important anti-inhibitors factors and are transmitted via the feed (Krogdahl, 1989). The presence of these antinutritional factors limit the use of vegetable feedstuffs in aquaculture. In practice, soybean remains the main ingredient from vegetable origin used in formulations because most of its ANF are heat labile and the processing of soybean is rather well controlled.

Toxicity of Additives

Certain feed additives are toxic when used at high concentrations. Errors in formulation or in ingredient proportionining can cause severe mortality in shrimp farming. Many commercial feed companies might have probably experienced these unfortunate situations. The implementation of strict production and control procedure to avoid this "human factor", is mandatory in feed preparation.

Fat Soluble Vitamins

Vitamins A, E and D are liposoluble and can be toxic when used at levels exceeding the requirements (Tacon, 1985; National Research Council, 1993).

1. Vitamin A can lower growth rate and cause anaemia at levels exceeding 100 times the requirements (1,000,000 IU/kg).
2. Vitamin D4 toxicity is probably linked with presence of calcium in water and in the feed. Feeding 50.000 U/kg Vitamin D3 (10–20 times the requirements) will depress growth and cause hypercalcemia.
3. High concentrations of vitamin E (5000 ppm) can cause reduced concentration of erythrocytes in blood and levels higher than 2500 ppm may depress the killing ability of macrophages (Lail and Olivier, 1993).

Trace Minerals

Although all trace minerals will be toxic when used at high concentrations, selenium is certainly the most toxic of all and is responsible for a number of disorder. The dietary requirements for selenium are in the order of 0.1 to 1 ppm. Although fish meal is a good source of selenium, almost all mineral complements do contain selenium, mostly as NaSe.

Dosage of higher than 10 ppm may cause mortality in most animals (Hamilton *et al.*, 1990) and levels of 50 ppm are considered to be lethal (Hilton *et al.*, 1980; Tacon, 1985).

Antibiotics

Antibiotics have been invaluable in the aquaculture industry. However, their use is now controversial because of the increased bacterial resistance it develops in animals (Brown, 1989; Dixon, 1990). Moreover, antibiotics like oxytetracycline can depress the immune system (Anonymous, 1994). The DNA synthesis of mitogen-stimulated pronephric cells was reduced to about 50 percent at therapeutic doses of oxytetracycline. Levels of 10 ppm were sufficient to reduce the activity of both the specific and non specific system in salmonids, whereas oxolinic acid did not cause this immuno-suppression (Siwicki *et al.*, 1989).

Horizontal Transmission of Pathogens through Feed

Transmission of diseases through the feed is a major concern for the farmers and the feed producers. For years now, *Salmonella* has been carefully monitored in feed ingredients to avoid contamination through the feed. Serious losses in poultry farming have been attributed to Salmonella infection. It has been shown recently (Mortesen, 1993) that infectious pancreatic necrosis virus (IPNV) can be transmitted to trout via feed pellets.

Some machine manufacturers are advertising special conditioners for pellet mills that kill *Salmonella* due to a long exposure of feed ingredients (30–60 seconds) to high temperatures (90–100°C). Extrusion technology ensures the production of a sterile feed, at least when the feed leaves the die of the extruder. Good sanitary procedures in the feed plant will limit the contamination during drying, cooling, coating, transport and packaging. Salmon feed manufacturers now avoid the use of salmon fish meal (by product from filleting industry) to reduce the risk of cross contamination.

More and more shrimp feed producers in South-East Asia and Latin America are now reconsidering the use of shrimp head meal in feed formulation since there is every possibility of transmitting viruses such as yellow head virus, White spot syndrome virus (WSSV) or Taura syndrome virus (TSV) through the feed via the use of processed shrimp head originating from contaminated shrimp ponds.

Micronutrients in Enhancing Immunity

Nutritional studies are assessing the requirement for a defined nutrient under optimal culture conditions. If feeds are formulated according to these data, it would underline that fish or shrimp would always be raised under the same optimal conditions in commercial operations. However, some problems may arise and place the animal in different situation (oxygen, temperature or crowding stress, presence of pathogens, etc.) in which the requirements for macro and micronutrients are likely to be different.

Several studies have been carried out for assessing the requirements for micronutrients under conditions such as high immunological activity, aggression by a pathogen or unfavourable environment conditions. It is essential for the nutritionist to provide these with consideration as the growth of shrimps are likely to be affected due to this problem.

Vitamin C

Vitamin C or ascorbic acid is the major water soluble antioxidant. Several roles are attributed to vitamin C and it acts as a biological reducing agent for hydrogen transport, in synergy with vitamin E and selenium to maintain activity of glutathione peroxidase and superoxide dismutase. Vitamin C is a necessary factor in the regulation of steroid and collagen synthesis (epithelial barrier). In homeotherms, it has been shown to be important in iron metabolism, antibody response and other immune functions. Phagocytic cells generate superoxide anions and hydrogen peroxide that might cause oxidative damage to the host and the white cells in general. Vitamin C protects the host against the damage as it would enhance phagocytic activity of macrophages.

Although the Vitamin C even at 10–20 ppm levels favour normal growth and surival, higher levels of vitamin C as known to improve immunological defenses, resistance to oxygen or salinity stress, etc. Thus, the requirement for vitamin C is highly varying and is depending on the on-farm conditions. Kontara *et al.* (1996) showed that 40 ppm of vitamin C was sufficient for optimal growth of *Penaeus monodon* postlarvae, whereas 2000 ppm gave 100 percent survival during the rearing period and when the animals were at salinity stress. Felix *et al.* (2003) confirmed that vitamin C induced immunity could be sustained in *P.monodon* and *F.indicus* for upto 12 days. This was demonstrated through ELISA based prophenoloxidase assay under laboratory conditions.

Dietary vitamin C levels lower than 100 ppm would avoid deficiencies, maximum growth at levels of 100–150 ppm and improved immunity at 2000–4000 ppm. However, it is known that vitamin C at 4000 ppm level would cost around 30 percent of the total ingredient cost of the feed and it will not be economical for shrimp culturists.

Vitamin E

Vitamin E is an essential fat soluble vitamin primarily functioning as an antioxidant. It is the major lipid soluble antioxidant in membranes where it protects unsaturated fatty acids. It has been shown to affect phagocytosis and humoral and cellular immune responses, and to enhance proliferation, chemotaxis and bactericidal activity of phagocytes. However, excessive dose may reduce intracellular killing ability if that killing depends on peroxidic damage to engulfed organisms. Vitamin E stimulates cell proliferation in immunopoietic organs, increases the number of antibody-producing plasma cells and stimulates T helper lymphocytes. It also modulates prostaglandin, thromboxane and leukotriene biosynthesis.

Vitamin A

Vitamin A is a fat soluble vitamin important in maintaining the integrity of epithelial and mucosal surfaces. Vitamin A influences in homeotherms haemotopoiesis of phagocytes and lymphocytes. Phagocytosis and intracellular killing may increased in shrimps which are fed with supplemental vitamin A. It restores normal humoral and cellular activity in steroid–treated animals. Retinoids and carotenoids increase the proliferation of T helper cells, the induction of cytotoxic T cells and expression of IL-2 receptors on natural killer cells. Vitamin A is a relatively weak anti-oxidant (Bendich, 1992; Blazer, 1992; Lail and Olivier, 1993).

Carotenoids

It has been illustrated that astaxanthin has an anti-cancer effect and increases the number of communicating cells. Carotenoids (b-carotene and canthaxanthin) are potent antioxidants and are known to enhance the proliferative responses of T and B lympthocytes to mitogens, increase cytotoxic T-cell and macrophage tumor-killing activities and stimulate the secretion of tumor necrosis factor and simultaneously lower the tumor burden. Canthaxanthin lacks vitamin A activity while beta-carotene is a precursor of vitamin A.

Tacon and Kurmaly (1996) reported that wild shrimp contain more astaxanthin (50–80 ppm) than cultured shrimps (10 ppm). Best growth rates have been obtained in *P. japonicus* fed with 200 ppm astaxanthin. However, use of astaxanthin in feed is restricted owing to its price.

Trace Minerals

In iron-deficient shrimps, antibody production and NK cell activity are severely impaired. Recently, the iron binding proteins transferring, lactoferrin and ferritin have been shown to affect immuno-regulatory properties. It is suggested that transferring binds circulating iron making it unavailable to the invading organism, hence overloading serum with iron may overwhelm the iron binding ability and increase susceptibility of the host. Iron also influences the production of interferon and prostaglandin. It has been reported recently that commercial feeds commonly contain 200–350 ppm of iron and that these levels may be detrimental to the health of aquatic animals (Blazer, 1992).

The multilevel defence system to protect cells against chain reaction initiated by free radicals comprises superoxide dismutases (SOD) that are Cu, Zn and Mn dependent. The activities of both the cytosolic Cu-Zn SOD and michondrial Mn SOD in the liver of fish were significantly altered in response to changes in dietary intake of these minerals (Cowery, 1986). Levels of 8 and 16 ppm of fluoride were effective in reducing mortalities (3–5 per cent) due to bacterial kidney disease in trout (Blazer, 1992).

Chapter 6
Probiotics and Immunostimulants in Shrimp Aquaculture

A. Probiotics

Probiotics are very innovative and a promising mechanism to control shrimp diseases. They have been widely and successfully used in agriculture and their use has resulted in a sharp decrease in antibiotic consumption (Fuller, 1989; Walter, 1990; Castaldo, 1991; Smith, 1991; Sainsbury, 1993). Studies have shown that the probiotic organism used in agriculture such as *Bacillus, Lactobacillus, Streptococcus*, etc. can also be useful in aquaculture (Gatesoupe *et al.*, 1989; Gatesoupe, 1991). However, as these organisms are not endemic to sea water they have only a regular basis to provide any beneficial influence. Most of the commercial products available today are for the aquaculture market, "repackaged" product's means for agriculture purpose. The current use of isolated marine bacteria in shrimp hatcheries in Ecuador is worth mentioning *V. alginolyticus* and other bacteria are nowadays selected from successful shrimp larval tanks and are mass cultured as probiotics (Austin *et al.*, 1995; Griffith, 1995). Their use resulted into a solution to the disease known as "las bolitas" caused by *V. parahaemolyticus*, thereby it has cut the use of antibiotics by 95 percent, increased production volume by 35 percent and reduced the shut down duration of hatcheries to seven days a year from 21 days.

Similar work was initiated on turbot (Westerdahl *et al.*, 1991; Olsson *et al.*, 1992) revealed that out of the 400 bacterial strains isolated from turbot gut, 89 were found to inhibit the growth of *V. anguillarum*. The application of these findings to commercial operations may be very useful for the shrimp aquaculture industry.

Probiotics are bacteria, microbes or cellular products are known to protect culturable species from pathogenic and infections (Jory, 1998), when they are added to culture media such as pond or tank water and/or when ingested by the cultivable species. Probiotics are beneficial to the target species in different ways. Probionts prevent infections by competitive exclusion, antimicrobial actions, or by some other means. Probionts are well recognized and used with terrestrial farm animals and humans (Morishita *et al.*, 1971; Muralidhara *et al.*, 1977; Sandine, 1977; Gilliland, 1979; Ellinger *et al.*, 1980). These applications included ingestion of *Lactobacillus acdiophilus* to prevent pathogenic, intestinal infections.

Application of probiotics to aquaculture mostly involve two approaches: (*i*) pond water additives and (*ii*) feed additives. Most of these applications have been developed during the last 10 years, with mixed results.

(i) Probiotics as Pond Water Additives

There is a variety of microbial-based pond "supplements" marketed today that are claimed to do many useful actions for shrimp and for shrimp pond water quality. Their beneficial mechanisms include improved water quality through accelerated mineralization and nitrification and/or improved cultivar health through ingestion of the probiont. These pond water treatments are commonly referred to as "bioaugmentation" (Moriarty, 1997). These products include bacterial innoculum, enzyme preparations, and plant extract products. The manufacturers claim that their products improve water quality, prevent off-flavor, reduce blue-green algal growth, accelerate sediment decomposition, improve cultivable health, and increased crop yields. Mostly, the manufacturers rely on testimonials from satisfied farmers, with little or no statistical evidence to back-up the benefits. Virtually all earlier studies show that there are no benefits from these open pond water additives (Tucker and Lloyd, 1985; Boyd *et al.*, 1994, 1995b, Boyd and Gross, 1999).

(ii) Probiotics as Feed or Feed Additives for Shrimp Larvae and PL

Bacterial probiotics have been used successfully with oysters, fish, and marine shrimp (Gatesoupe, 1991; Maeda and Liao, 1992; Douillet and Langdon, 1994). In each case, the bacteria of probiotics were ingested by the cultivar species and as a result its survival and/or growth was improved. The following examples demonstrate two successful applications with marine shrimps.

Garriques and Arevalo (1995) isolated non-pathogenic *Vibrio alginolyticus* from raw seawater in Ecuador. *Vibrio alginolyticus* was then batch cultured and added to *P. vannamei* larviculture tanks (25 and 60/m^3). Control larviculture tanks containing oxytetracycline additions or with no prophylactic additions were used for comparison. Average survival from the probiotic treatment was 90 per cent, compared to 83 and 74 per cent for antibiotic and no treatment tanks, respectively. Water samples from larviculture tanks indicated no pathogenic *V. parahaemolyticus* in tanks receiving the *V. alginolyticus* probiotic, while the other two larviculture tanks contained *V. parahaemolyticus* at 10 per cent of the water samples. These findings clearly indicate that probiotic bacteria could provide protection against infections by pathogenic bacteria during larviculture.

Rengpipat *et al.*, (1998) demonstrated similar beneficial effects with tiger prawns (*P. monodon*). They isolated a species of *Bacillus* (designated as S11) from marine sediments, ocean waters, and from *P. monodon*. This bacterium showed antimicrobial properties in all innocula. It was fed to PL and juvenile shrimps in a prepared diet in three forms, fresh, in saline solution, and freeze-dried. Compared with controls, all the three forms recorded a significantly greater ($P < 0.05$ per cent) survival and growth of *P. monodon* during 100 days culture starting from PL–15. Furthermore, shrimps fed with *Bacillus* S11–fortified feeds had significantly greater ($P < 0.05$) survival compared to controls when challenged with pathogenic, luminescent bacterium *Vibrio harveyi*. During these 10 day challenge tests using 115 day-old *P. monodon*, survival was 100 per cent with shrimps fed with Bacillus S11 in all three forms, while survival was only 26 per cent for shrimps not fed with the probiotic. Similar benefits were recorded when *Bacillus* S11 was fed to *P. monodon* during larviculture, although survival rate during challenge tests with *V. harveyi* was much less for both control and probiotic treated larvae (Rengpipat *et al.*, 1999). Tsumura (1999) also found that bacterium coded BY-9 when fed to the larvae of *P. monodon*. They found 46 per cent survival to PL stage compared with 11 per cent survival for non-inoculated controls.

B. Immunostimulants

Immune defence mechanisms in penaeid shrimps exist, but they are not well developed compared to vertebrate animals (Vargas-Albores *et al.*, 1998). Shrimp immune responses are more short-term and lack broad spectrum. One application of vaccines does not provide long-term protection to shrimp. A more appropriate approach with shrimp is the continuous use of chemicals and microbial products that stimulate the shrimp's immune response against potential pathogens. In particular, peptidoglycans, lipopolysaccharides, and β-glucans found on the surfaces of bacteria and fungi can cause immune system stimulation in penaeid shrimps when administered in feed.

Several laboratory studies have demonstrated immunostimulation by diet supplements. When heat killed *Vibrio harveyi* bacterium, and yeast glucan were fed to *P. monodon* they caused peak immune responses after 48 h (Devaraja *et al.*, 1998). Fucoidan, a sulfated polysaccaride extracted from *brown algae*, protected *P. japonicus* from

WSSV challenges (Takahashi *et al.*, 1998). Shrimp fed with fucoidan had greater survival (80 per cent) during a 10-day challenge test with WSSV, while control shrimp had nil survival. Itami *et al.* (1998) observed even greater protection with *P. japonicus* fed peptidoglycan derived from the bacterium *Bifodobacterium thermophilim*. Shrimps fed with this immunostimulant should have had more than 90 per cent survival during 40 to 50 days when challenged with WSSV, but their survival decreased to zero per cent within 10 to 20 days after the immunostimulant was withheld. Some of these immunostimulants, such as fucoidan are presently too expensive for use in commercial farming operations.

By definition, an immunostimulant is a chemical, drug, stressor, or action that elevates the non-specific defence mechanism or the specific immune response (Anderson, 1992). This definition describes the complexity of the interaction between the immune system and its environment. The immune system can be found in any possible state between the normal state and the high "alert" state. Each individual is also likely to present a different state. Any environmental stress (Fevolden *et al.*, 1994) of any unusual substance present in the food may have an effect. This also explains the wide variety of immunostimulants in use today (see the Table 8) (Anderson, 1992; Secombes, 1994).

The sole injection of a phosphate buffer will already trigger some defence mechanisms. On the other extreme, the injection of a vaccine or bacterin together with an immunostimulant will probably give the highest response. A number of studies have dealt with the evaluation of well known immunostimulants such as polysaccharides, lipoproteins, lipopolysaccharides, etc. mainly derived from the cell wall of microorganisms.

Yeast

For a long time yeast has been known for its effect on health and disease resistance. Thus yeast was often fed to exotic animals before or after their transfer to zoos to decrease the mortality caused by the transport and the related adverse conditions. Experimental studies with monkeys showed that the oral administration of brewer's yeast could increase their resistance to infection (Sinai *et al.,* 1974). A two weeks lag was necessary between the initiation of the yeast treatment and the expression of the resistance. The effect was attributed to the *in vivo* stimulation of phagocytosis. Yeast is also favoured by feed formulators since it is cheap and yeast with B vitamins, is a good protein source and is an anti-stress factor. It is often used as a cheap additive (when sales prices permit its inclusion), as the " magic touch " of the formulator or as a carrier for some premixes.

Glucans

The substance, glucans have probably attracted the attention of researchers in aquaculture more than any other immunostimulants. Their availability and "reasonable" price (still 8–16 per cent of total ingredient cost) is probably a reason for that situation. Glucans, as all immunostimulants, are used either by direct injection into the body (parenteral injection) or by oral administration via the food (per os). The results obtained by one way of administration can not be extrapolated to the other. Although the effect of glucans via parenteral injection is well described (Yano *et al.*, 1989; Chen and Ainsworth, 1992; Jergensen *et al.*, 1993; Rerstad *et al.*, 1993) and often used to potentiate the reaction on vaccination, the effect of glucans per os is still controversial. An aminated β 1-3 polygiucose (3H–labelled) has been tested on Atlantic salmon (Ingebrigtsen *et al.*, 1993) via parenteral injection and oral administration. Parenteral injection resulted in high levels of radioactivity in organs with large amounts of lymphoid tissues, such as spleen, pronephros, gills and wall of the posterior intestine and the radioactivity persisted at high level through out the experimental period (96 h). However, substantial levels of radio-labelled substances following the oral administration was only located in the wall of the posterior intestine, whereas only traces were found in other tissues.

Commercial Immunostimulants–Key Issues

An immunostimulant is nothing but an antigen that stimulates the immune system. It is therefore incorrect to expect from immunostimulants a complete protection against pathogens and adverse conditions. Survival is the final result of a long list of events and a stimulated strong immune system is just one step in the right direction. The problem is indeed not simple and the immunostimulant product range is still complex, as we can see from the Table 8.

There are obviously some products which work better than others. For example, differences between various types of glucans can be related to their tertiary structure which has to be recognised by the receptor present on the macrophage. Other product like mannans which are supposed to flocculate micro-organisms within the lumen of the digestive system, should be evaluated for their selectivity. One should also consider that a feed pellet, in essence, already contains antigen like small peptides, nucleotides, complex carbohydrates originating from ingredients like fish meal, plant proteins, cereals, etc. How far these "natural" antigens might already doing part of the job is still an open question. The choice of one product unfortunately is not an easy task and the motivation for selecting one product or another is often strongly subjective.

As described above, the benefit of oral administration of immunostimulants is still unclear because their absorption by the blood is minimal. Lots of products will work well *in vitro* on isolated organs or when directly injected in the blood but will fail to work properly with given per os. That limits their effect on the immune system.

Careful Experimental Designs and Analysis

Most immunostimulants will indeed activate the macrophages *in vitro*. Even a phosphate buffer solution could do the job (Anderson, 1992). But an experiment *in vitro* does not confirm if the product will effectively stimulate the immune system of the animal when give orally.

Another characteristic of these experiments (Devresse *et al.*, 1997) is the very high variability observed. The variability between individuals is often higher than the variability between treatments, giving statistically non significant results. Moreover, there is not a good correlation existing between the cellular or humoral response of the immune system and the survival at the end of the experiment.

Externally stressing the animal (manipulation, low oxygen or high metabolites concentration, light, pH) is often having a more pronounced effect on the immune system than any immunostimulants (Devresse *et al.*, 1997). This is clearly seen under lab conditions on blood parameters such as haemotocrit and leucocrit levels, lysosyme activity or NBT values.

Although challenge tests are one of the best criteria to evaluate a product within a laboratory, they are not a good indicator on how the product will perform on the field. The level of pathogens injected to an animal during a challenge is carefully chosen to disclose a difference between treatments and a higher concentration would show a high mortality and again no significant differences. Under practical conditions, in a fish pen or a shrimp pond, the quantity of pathogens attacking the animal is not under control and might not be the "right number" needed to show the positive effect of the activation of the immune system. The farmer will then see no advantages in the use of immunostimulants.

In practice however, immunostimulants are still an interesting aspect to be used in animal feeding to enhance disease resistance, although the effect is not clearly and repetitively seen each time immunostimulants are fed. Immunostimulants are one of the tools that the farmer or the feed manufacturer can use to fight the diseases. They should be used together with optimal and healthy rearing conditions, adequate nutrition, good and strong shrimp post-larvae, etc.

The best products are not necessarily the most expensive ones. When feed cost is a limiting issue, yeast should still be considered as an additive of choice. There is a wide variety of yeast products available on the market namely torula yeast, molasses yeast, brewer's yeast which are being sold plasmolysed or not, drum dried, spray dried, whole or after as extraction (cell walls). Their inclusion is not expensive and their effect is well documented.

Table 8: Immunostimulants, Adjuvants and Vaccine Carriers Used Routinely and Experimentally in Fish and Other Animals (Anderson, 1992)

a.	**Reservoirs and depots**
	Freund's adjuvant
	Liposomes
	Mineral oil
	Lanolin
	Parafin oil
	Dextran sulfate
	Ethylene-vinyl acetate
b.	**Carriers and vehicles**
	Bentonite
	Kaolin
	Latex beads
	Sheep red blood cells
	Dimethylsulfoxide
c.	**Inflammatory agents**
	Silica particles
	Carbon particles
d.	**T cell stimulators**
	Mycobacterium sp.
	Muramy! dipeptide
	Glucans (yeast extracts)
	Metallic salts
	Levamisole
	Bacille calmette guerin
	Corynebacterium parvum
	Polynucleotides

Contd...

Table 8–Contd..

e.	**B cell stimulators**
	Lippolysaccharides
f.	**Cell membrane modifiers**
	Detergents
	Sodium dodecy! sulfate
	Quaternary ammonium compounds
	Saponins (plant extracts): quil
g.	**Animal extracts**
	Ecteinascidia turbinata extract
	Haliotis discus extract
	Fish extract
h.	**Mitogens**
	Pokeweed mitogen
	Phytohemagglutinin
	Concanavalin A
i.	**Nutritional factors**
	Vitamin C
	Vitamin E
j.	**Cytokines**
	Interleukins (leukotriene B4, interferon)
k.	**Heavy metals**
	Cadmium
	Germanium

We see no technically established proof that probiotic pond supplements benefit shrimp grow out. Probiotic feed supplements, however, have clearly defined benefits during larviculture and early rearing. What is needed now are grow out trials in pond settings to document potential benefits of these feed supplements. In particular, we need to know: (i) Will probiotic uses during larviculture confer protection through growout, or should the probiotic be added to feeds during part or all of growout? (*ii*) Will probiotic bacteria protect against viral pathogens ? and (*iii*) Are there other probionts that would be more effective against bacterial and/or viral pathogens ? We can expect commercial applications of probiotic feed additives to be used widely and significant improvements in their efficacy in the near future.

Immunostimulant feed additives are now being used in shrimp culture systems, and research results look very promising for further improvements in pathogen protection. Improved protection and cost effectiveness should result in more widespread use of immunostimulants in shrimp culture systems.

Chapter 7
Shrimp Raceways: Operation and Management

For design and construction details for raceways, refer Chapter 1.

Preparation of Raceways for stocking

(i) Filling

As water quality parameters, algal species and dynamics are site dependent, the following guidelines need to be followed. Specific modifications will be needed for each site location.

Raceways will be filled with filtered seawater to working level 4 to 6 d prior to scheduled stocking. A filter pipe with a 600 µm screen should be mounted on the drainage outlet prior to stocking. Airlift pumps and air diffusers should be turned on to high flow capacity prior to any fertilizer application.

(ii) Fertilization

Based upon the prevailing temperature and algal abundance, raceways should be fertilized 4 to 6 d before scheduled stocking. Application of 5 ppm N, 0.5 ppm P, and 1 ppm Si has been found to enhance adequate algal bloom. However, this rate needs to be standardised for each location. To avoid precipitation, each fertilizer should be dissolved in water and distributed separately.

(iii) Counting and Storage

Assessment of the total number of shrimp nauplii should be made by taking samples from the concentrated nauplii. A sample of 1 ml should be taken after thorough mixing. This is diluted with 999 ml of fresh water. The total number of nauplii harvested should be determined from five 10 ml sub samples taken after thorough mixing of the 1000 ml sample. Nauplii counting can be done against a light source with or without a dissecting scope. The 1 and 10 ml samples can be taken by an ordinary pipette or Hensen-Stempel pipette. Another useful method is to pour each of the 10 ml samples into a Petri dish and count the nauplii against a black background. Nauplii density per litre should be marked on each storage vessel used, along with the date and time of harvest. Storage vessels with the nauplii should be transferred into a cold storage room of about 4°C with aeration. For best results, nauplii density should not exceed 3 million/l.

(iv) Preacclimation

Acclimation tanks should be cleaned and disinfected a few days before stocking of post-larvae. On the day of stocking, tanks should be filled with filtered seawater from the same raceway in which the post-larvae are to be stocked. Water temperature and salinity should be adjusted according to the anticipated salinity and temperature in which the post-larvae are shipped. The water quality instruments *viz.* salinity, refractometer, oxygen meter, pH meter and thermo-alignment metre should be checked and calibrated to ensure correct readings.

(v) Acclimation Process of Post-Larvae

In order to reduce acclimation time, it is recommended that salinity adjustment to be commenced in the larval-rearing tank prior to the harvest. However, if such acclimation is not possible, all efforts should be made to ensure smooth acclimation at the nursery site. Temperature, dissolved oxygen and water salinity values in the acclimation tank should be similar to the anticipated values of medium in the shipping container. Since a good algal bloom in the raceway water is expected, special care should be taken to ensure that water pH does not exceed 8.6. To reduce stress, post-larvae should be unloaded at the site in the early morning or late afternoon during the cool hours of the day. The salinity, temperature, dissolved oxygen, and pH should be measured upon arrival of these larvae at the nursery site. If differences in water quality between acclimation and hauling tanks are small, postlarvae can be transferred by gravity flow into the acclimation tank. If differences in water quality parameters between the two tanks are significant, gradual adjustment of the acclimation tank water is needed. For transport in polyethylene bags, all boxes should be selected randomly for measurement of salinity, temperature, dissolved oxygen, and pH. If differences in water quality between shipping and acclimation water are significant, gradual adjustment is needed. After opening the bags arriving post larvae, about 2 litres of acclimation tank water can be added and the bags should be left to float in the acclimation tank. Adding water to the bags should be done in increments of 2 l until differences between the two water sources are minimal. At that time, post-larvae bags should be above saturation, it is recommended that this high level be maintained in the acclimation tank water as well. Switching from oxygen to air bubbling should be done about 15 min after release of the last batch of postlarvae into the acclimation tanks.

During the acclimation process a stocking density of 500 to 1000 post-larvae per liter can be maintained. Post-larval acclimation should be done gradually by adding raceway water to the acclimation tank, using the raceway pump. Salinity, temperature and pH in the acclimation tank should be measured and recorded every 15 min. At the end of the acclimation process, post larvae can be released by gravity into the raceways. Certain guidelines relating to time needed for adjusting water quality parameters are given in the tables.

Adequate mixing of the post larvae is also needed so that post-larvae are randomly distributed at the time of sampling. Mixing can be done manually, or preferably by an aeration grid as described earlier. Fifteen samples (100 to 200 ml each) should be taken to obtain an accurate population estimation.

Sample volume is largely dependent on tank volume, as well as post larval density. Generally speaking, a sample size of 150 to 200 ml should be adequate for a 1000 l acclimation working volume. Samples should be transferred from the sampling container into a larger container for counting. If the post-larval counts from 10 samples are similar there will be no need to count all 15 samples. However, if variation is high, all the 15 samples should be counted and extreme values should be omitted in the final tabulation. The estimated number of post-larvae in the acclimation tank can be determined according to sample size and acclimation tank water volume. If the acclimation process is scheduled to take more than 4 h or if post-larval cannibalism is observed, immediate feeding is needed. About five *Artemia* nauplii may be required at the first feeding of each post-larvae. Additional feeding should be provided according to post larval consumption.

If the salinity, temperature, and pH parameters in the acclimation tank are in equilibrium with the raceway water, post larvae can be released into the raceway. Raceway stocking densities are largely dependent upon the system-carrying capacity. A biomass load of about 1.9 kg/m^2 has already been demonstrated for the nursery raceways even without oxygen injection or bottom manifold support. The carrying capacity for this suggested raceway system is approximately 2 kg/m^2.

(vi) Feed and Feed Ration

Feeding with dry feed should be started on the first day of stocking. Post-larvae should be fed six times a day with two ration size, (Table 4). The post larvae should be fed with dry feeds containing 45 to 55 per cent crude protein. A simple electronic spreadsheet (such as Lotus 1-2-3, Lotus Development Corp., Cambridge, MA) can be used to create a table which will automatically recalculate the ration size according to animal growth and expected survival. The feed should be provided in different particle. The suitable size ranges of the feed particles commonly used for different post larval size groups are shown in Table 5.

Table 1: Salinity Acclimation Schedule for *Penaeus sp.* PL_{1-5} in raceways

Salinity Range	*Min Needed for 1 ppt Change*
45–50	60
40–45	30
32–40	15
32–22	15
22–16	30
16–8	60

Note: Optimal salinity for nursery is < 31 ppt.

Table 2: Temperature Acclimation Schedule for *Penaeus* sp. PL_{1-5}

Temperature Range (°C)	*Time Needed for 1°C Change (min)*
18–25	15
25–30	30
30–32	60

Table 3: pH Acclimation Schedule for *Penaeus* sp. PL_{1-5}

pH Range	*Time Needed for 0.1 pH Unit Change (min)*
6.0–8. 2	Direct transfer
8.2–9. 0	15
9.0–9.7	30
9.7–10.0[a]	60

a: pH values over 9.7 are not recommended and should be avoided.

Table 4: Ration Size and Feeding Schedule for *Penaeus* sp. PLs

Feeding time	3:30	8:30	11:30	14:30	17:30	22:30
Ration size (% of daily ration)	21	12. 3	12. 3	12. 3	21	21

Table 5: Feed Particle Size and PL Mean Weight

Feed Category	*Crumble 0*	*Crumble 1*	*Crumble 2*	*Crumble 3*
Particle size range (mm)	< 0.59	0.59–<1	1–<1. 41	1. 41–< 1. 68
Postlarval average weight (mg)	1–15	15–50	50–500	> 500

For the first week, post larvae are only fed crumble size 0. Starting with the second week, the postlarvae diet should contain 25 to 50 per cent crumble 1 and rest with crumble 0. During the third and fourth weeks, the diet composition should contain three components *viz.* crumble 0, 1, and 2. The proportion of each feed type should be determined by the size variation in the population. If post-larvae smaller than 15 mg are not common among the sampled population, the crumble 0 diet can be omitted. The daily amount of feed should be determined primarily according to post-larval feed consumption. Observation of feed consumption can be accomplished with an underwater viewing apparatus along with bottom sampling with a fine-mesh dip net. Feeding level may be much higher if natural productivity is low.

(vii) Supplemental Feeding

From the second week, supplemental feeding with live or frozen adult *Artemia* is highly recommended. The number of supplemental feedings per week should be determined by the health condition of postlarvae or fouling level. If heavy fouling or surface bacterial contamination is detected, frequent *Artemia* feeding should be considered to enhance molting. Under a low incidence of integument fouling and good animal health condition, supplemental feeding should be given only twice a week. Nevertheless, the total amount of *Artemia* given during one cycle should not exceed 10 per cent of the total expected shrimp biomass at harvest.

The feed ration records should be updated on a daily basis. These data should be used for calculating the feed conversion ratio (FCR) as well as for evaluating post larval growth performance. 'Data recording sheet' is a suggested format which can be used for recording feeding data.

(viii) Managing Water Quality Parameters

Monitoring water quality parameters is a key factor for successful nursery operation. Close observation of these parameters allows maintaining optimal growing conditions.

(a) Temperature

Temperature should be measured twice daily, in the morning and in the afternoon. High water temperature (above optimal) increases stress and reduces growth which may ultimately lead to poor survival. For optimal growth, the raceway water temperature should be maintained at 26 to 30°C. If water temperature is more than 31°C, measures should be taken to reduce the temperature.

(b) Dissolved Oxygen

If a dissolved oxygen monitoring system with automatic oxygen injection is employed, daily checking of the sensors and injection system is needed. If oxygen injection is manually controlled, the dissolved oxygen level should be measured at least twice a day: in the morning and in the afternoon. From the third week on wards when the raceway biomass is high, measurement of dissolved oxygen at 22:00 hrs and 4:00 hrs should be performed to ensure an adequate dissolved oxygen level. Metre calibration should be done according to the manufacturer's specifications to avoid misreading. The oxygen injection system should be switched on any time the dissolved oxygen level decreases below 4 ppm. Ideally, the oxygen injection rate should be regulated to maintain the dissolved oxygen level above 5 ppm.

(c) Secchi Disc Visibility

Secchi disc visibility readings should be taken twice daily. If the sediment load of intake water is minimal, secchi disc readings provide a good estimate of the phytoplankton population in raceway water. The optimal secchi visibility reading is different for each location as well as for each growing season. Salinity, temperature, light intensity, nutrient level, natural algal abundance, etc. can enhance blooms of different algal species. If the algal bloom is of a desirable species and the pH is adequate, a secchi disc visibility of 25 to 30 cm should be maintained for at least the first 2 weeks of growth. If the algal species, pH, or other water quality parameters are not optimal, secchi visibility should be adjusted on need basis.

(d) Salinity

P. monodon postlarvae can grow in a wide salinity of range. Preferably, a salinity of 25 to 30 ppt should be maintained through out the nursery phase. To avoid wide salinity changes, it should be measured twice a day in all raceways as well as in the reservoir pond (For salinity acclimation Table 9 may be referred..

(e) Algal Cell Counts

Algal dynamics of the sea water can be site specific and algal counts may be done on a daily basis during the first year of operation. Estimating algal bloom and condition can be done by a blood cell counting chamber (Hemocytometer). This counting chamber has two grids, each with four squares in each corner ("L" squares, Figure 1) and one square at the center ("D" square). For

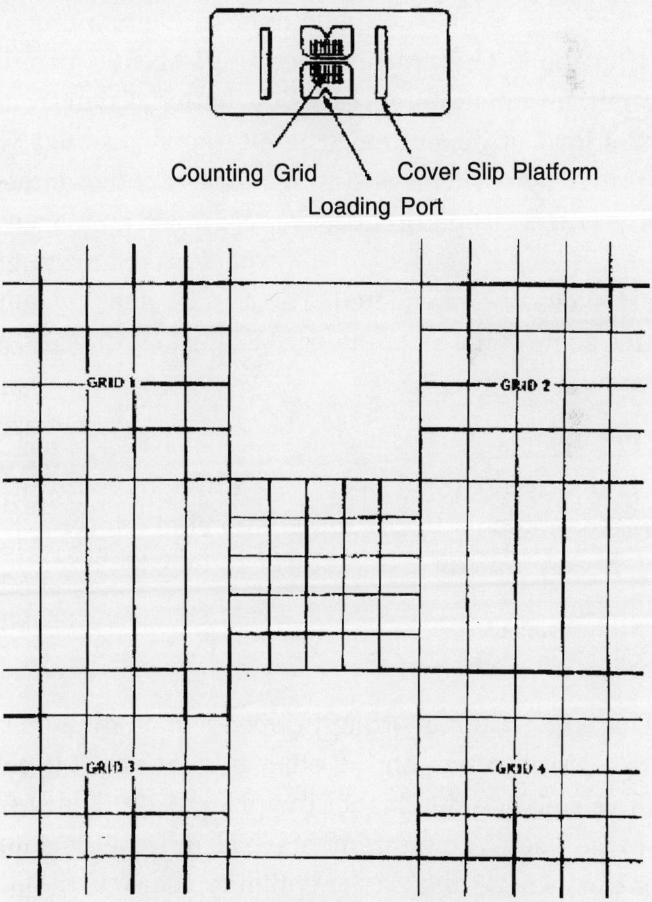

Figure 1: Algae Counting Chamber

estimating algal densities, only cells located inside the "L" and "D" squares are counted. Note that although the "L" and "D" squares have different divisions, their total areas are equal. The total number of cells per milliliter is obtained from the product of the average number of cells per "L" square by 10^4. Algal counts are to be estimated for both the reservoir and the raceways.

(f) pH

Estimation of pH should be done in the morning and the afternoon. The morning readings should provide important information regarding the bacterial activity and biomass load of the system. The afternoon readings will ensure that high pH values–due to intensive photosynthesis activity–will not jeopardize post larval growth and survival. If high pH levels (above 9.6) are experienced, lowering of the pH should be considered. This can be done by either flushing some of the algae out or injecting CO_2 directly into the raceway water.

(g) Ammonia

Ammonia nitrogen occurs in water in two forms, unionized ammonia (NH_3) and ammonium ion (NH_4^+), in a pH-, temperature-, and ionic strength-dependent equilibrium:

$$NH_3 + H_2O \rightarrow NH_4^+ + OH^-$$

This equilibrium is strongly dependent on solution pH and to a lesser extent, on the other two factors. It is clear from this equation that higher the pH, greater the level of NH_3 in the water. A high ammonia (NH_3–N) level in culture media can increase disease susceptibility, damages the gills, and reduces blood oxygen transport ability. Shrimp tolerance to ammonia varies with species, physiological condition, and environmental factors. Lethal levels for juvenile shrimp for short-term exposures (48 to 96 h) are between 0.4 and 2.31 mg/l of NH_3. The Wickins "maximum acceptable level", which also reduces growth by 1 to 2 per cent of that of controls, was 0.1 mg/l of NH_3–N (corresponds to 1.8 mg/l of NH_4^+ -N at 28° C, pH 8.0, and salinity of 27 ppt). Since most algae can utilize ammonia, the unionized ammonia level should not be a problem when a good algal bloom exists in the raceway water. Therefore with a good algal bloom, ammonia should be measured only twice a week. In the case of algal crash, daily monitoring of ammonia levels is needed.

Ammonia can be measured by ammonia electrode, commercial kits, or by an analytical wet chemistry procedure.

(h) Nitrite

The tolerance of aquatic organisms to nitrite (NO_2) is considered to be high (the 96 h LC_{50} value range is from 8.5 to 15.4 mg/l). With adequate water management, all waste materials may be removed from the raceway. Normally, the nitrite level in the raceway water should be very low. Nevertheless, as a precaution, nitrite should be measured twice a week to ensure that the level is below 1 mg/l. The nitrite level in the water can be determined by electrode, commercial test kit, or analytical method.

(i) Nitrate

Nitrate (NO_3) is the end product of ammonia oxidation performed by bacteria. The toxicity of nitrate to shrimp is considered to be very low, and only very high concentrations of nitrate are believed to have adverse effects on shrimp.

The 96 h LC_{50} nitrate to juvenile shrimp is 3400 mg/l. As high nitrate levels in raceway water are not normally anticipated, nitrate level is not to be estimated routinely.

(j) Hydrogen Sulphide

Under anaerobic conditions, certain heterotrophic bacteria can use sulfate and excrete sulphide. The un-ionized hydrogen sulphide is toxic to aquatic organisms. Any detectable concentration of hydrogen sulphide is considered undesirable, since a concentration of 0.01 to 0.05 mg/l of H_2S may be lethal to aquatic organisms. The presence of hydrogen sulphide is easily recognized without water analysis, since the "rotten-egg" smell of hydrogen sulphide is detectable even at low concentrations. Avoidance of an anoxic condition in raceways is very important for successful nursery operation.

(ix) Water Exchange

A diatom bloom has been observed to be beneficial for post-larval growth and survival in an intensive nursery system. Enhancement of algal blooms prior to post larval stocking is recommended, provided the desired algal species is present, and the algal bloom should be maintained as long as possible. The water exchange rate during the first 2 to 3 weeks can be limited, due to the low biomass and good algal bloom. The water exchange rate is normally increased in relation to increase in shrimp biomass. These values are valid for a minimal average water temperature of 26° C and a final harvest biomass of about 2 kg/m^2. The suggested exchange rate is subject to changes according to water quality criteria and post larval health condition.

To resuspend settled particulate matter, 1 h before the scheduled water exchange, the bottom manifold should be engaged and the airlift pumps set to full pumping rate. The strong current created should help move the particulate matter towards the sump. After 1 h, the external standpipe should be used to drain the predetermined volume of water. At the end of the water exchange, the bottom manifold is disengaged and the airlift pumping rate is adjusted to the normal level.

(x) Filter Screen Size

Filter screen size on the outlet filter pipe is an important parameter in raceway water management strategy. By choosing the right screen size on the outlet filter pipe, fecal matter, leftover feed, and particulate matter can be removed from the system. Filter screen size should be adjusted according to the size of post larvae to avoid escape of the small size postlarvae. Table 6 provides a tentative schedule for replacing filter screens during a 5 week nursery run. In a population with high size variation, some modifications are needed. Increasing the filter size should be postponed until the smallest post-larvae cannot pass through the larger–size screen holes.

Table 6: Weekly Ration Size for 5 Week Nursery Period with *Penaeus* sp. PLs

Feeding Level	*Week*				
	1	*2*	*3*	*4*	*5*
Daily feeding level (per cent of total estimated biomass)	10	10–15	15	15–20	10–15

(xi) Bottom Condition Evaluation

The bottom assessment of raceways should be done every morning before feeding. Three methods are suggested for this evaluation: (i) checking the bottom with an underwater viewing apparatus, (ii) observing the water column while the bottom manifold is engaged, and (iii) checking bottom samples collected with a dip net.

Table 7: Expected Average Daily Water Exchange Rate for Intensive Nursery Raceway System

Water Exchange Rate	*Week*				
	1	*2*	*3*	*4*	*5*
Average daily exchange (per cent)	10	30	30–50	80–150	200–300

Table 8: Recommended Filter Screen Size for Intensive Nursery Raceway System

Filter Screen	*Week*				
	1	*2*	*3*	*4*	*5*
Screen size (mm)	0.4–0.6	0.8	0.8–1	1	1–2

The first method is useful for observing dead shrimp, bottom stagnation, feed leftover, and feeding activity. The second method is especially useful during the first 2 weeks after stocking. Since postlarvae are small, they will be resuspended by the bottom current, which will facilitate the evaluation. The third method serves for feed consumption estimation.

All hydrological records along with the water management implemented should be recorded on a single data sheet. To avoid misinterpretation, data should be recorded with the time the measurement was performed. Daily Recording Sheet provides a suggested format for recording hydrological parameters.

(xii) Daily Assessment of Population

(a) Sample Size and Sample Preparation

To optimize raceway management, health of postlarvae should be evaluated on a daily basis. The evaluation should include information on deformities, activity, gut content, fouling level, molting rate, and animal stress signs in post-larvae. About 50 to 100 post-larvae should be immediately examined for gut content, swimming activity and stress signs. A thorough post-larval evaluation should be done by dissecting microscope. The post-larvae should be first observed under low magnification. Post-larvae need to be observed for deformities, abnormal swimming activity, hepatopancreas colour and volume, gut fullness, amount of fecal strands, molt abundance, mortality, predation, or any other unusual behaviour. Randomly selected 20 to 30 specimens are placed in individual drops of water for further evaluation.

(b) Disease Assessment

Assessment for the presence of disease should be done on at least 10 of the 20 to 30 post-larvae. Each one of the 10 specimens should be examined closely. It is not uncommon to find animals which are infected by several diseases or with several stress signs. Keeping health records for each examined post-larvae should provide an adequate population health evaluation. One suggested method is to assign a number from one to ten for each examined post larva. The assigned number will be marked under the corresponding infection/stress category and the affected organs. For example, if a post-larvae is infected with chitinoclastic bacteria, benthic algae and filamentous bacteria on the swimming legs and eyestalk the same will be marked under each of the infection categories in the corresponding affecting organ cells (*e.g.*, swimming legs and eyestalk). The collected data should be summarized by overall population infection level and the percent infection for each of the criteria. This information should help the raceway manager in the decision making process.

(c) Monitoring of Population Growth

Growth rate is an important parameter for evaluating post-larvae's performance in the nursery raceway system. The monitoring of population growth is done twice a week in order to obtain accurate growth data for determining feed levels and population health. To determine growth, a random sample of about 200 post-larvae should be collected from each raceway. Samples are taken from different sites in the raceway. During the first 2 weeks, a dip net with a fine-mesh screen of 800 μm can be used. A dip net with a 1.5 mm mesh size well be needed when the mean weight of shrimps is more than 50 mg. When sampling, the large postlarvae usually try to avoid the dip net; therefore, adjusting the dip net screen and frame size is needed to prevent biased sampling.

To ease the sampling process, post-larvae could be concentrated in the dip net and blotted dry on a paper towel for 30 sec. The biomass and total number of shrimps in the sample are recorded to calculate the mean weight of postlarvae. During the counting, small post-larvae should be separated and their biomass recorded. Estimating the size and relative percent of this group among the population is necessary to determine the outlet filter screen size and the feed particle sizes and proportion in the daily ration.

(xiii) Health Management

The rapid growth of the shrimp culture industry has resulted in an increase in disease with significant reduction in shrimp yield all over the world. An example is the collapse of the shrimp culture industry in Taiwan in 1987. The annual production dropped from 80,000 t in 1987 to only 20,000 t the following year. Similarly in India between 1994 and 1997, the sector faced unprecedented damage due to the outbreak of WSSV. Therefore, it is obvious that careful planning and avoidance of diseases are crucial factors for successful shrimp culture operation. When considering the occurrence of disease and parasites in shrimp nursery facilities, one should always bear in mind that water quality, nutrition, and pathogens are closely related to each other. An outbreak of an infectious disease may have been caused by poor water quality or in adequate diet, and not necessarily by poor-quality seed stock. Therefore, if post-larval quality is good, providing the post-larvae with adequate water and feed quality should result in excellent growth and survival. It is recommended that the use of water and feeding management strategies as primary tools for disease rather than heavy use of chemicals. However, it is to be noted that shrimps can be infected by a large number of diseases caused by viruses, bacteria, fungi and protozoa and also due to nutritional or environmental deficiencies.

(xiv) Harvest

(a) Hauling/Harvest Tank Preparation

Stocking density is crucial for successful pond management and hence shrimp biomass and average weight must be recorded at harvest. A few days before the scheduled harvest, hauling tanks and harvest equipment should be checked and cleaned. Sufficient oxygen cylinders should be available to avoid a shortage during transport. To save time during the weighting process, all harvest baskets should have an identical weight. Water temperature in the hauling tanks should be maintained at the same level as in the raceway. Water in hauling tanks should be enriched with oxygen during transport to avoid oxygen depletion.

(b) Weighing Station

An electronic balance with a bottom-weighing hook is recommended for the nursery harvest. The balance should be placed on a high table or a tripod to ease changing harvest baskets below the balance.

(c) Raceway Preparation

Before the scheduled harvest, the water level in the raceway should be decreased by three fourths every morning to flush the particulate load, and bottom manifolds should be operated for longer periods. On the day of harvest, the water level should be reduced by two thirds, the oxygen injection system should be switched on, and the raceways should be kept on flow-through. Raceways are ready for harvest when the water level is one third of the working volume.

(d) Harvesting and Data Records

The harvest basket should be checked periodically to avoid overflow and overcrowding conditions. It is recommended that the biomass load in the harvest basket is not exceeded 5 kg to avoid damaging the harvested juvenile shrimps. Only one person should be in charge of the data recording. This person should record the harvested biomass and keep track of the biomass in each of the hauling tanks. A second person is needed for the weighing process, a third person for carrying the harvest baskets to the weighing station, and a fourth person for carrying the harvest baskets to the hauling tanks. A fifth person is needed for measuring the oxygen and temperature in the hauling tanks, and he should also be responsible for washing the remaining postlarvae into the drain outlet.

The harvest basket should be left to drip for 30 sec. to remove excess water before reading the weight. For accurate survival estimation and to compensate excess water, deduction of an additional 5 to 10 per cent of the total harvested biomass is recommended.

(e) Population Sampling

If population size variation is small, group weight can be used to calculate a "crude" post-larval mean weight. Five to ten groups, each with about 200 specimens, should be taken randomly from several harvest baskets. The average weight should be determined according to the procedure described for the growth monitoring process. This average weight can be used for a "rough" pond stocking estimation. Individual weight records from random samples of 100 post-larvae should also be taken for the determination of population size.

(f) Transportation of PL

A hauling tank biomass of 100 g/l can be maintained for short trips without an adverse effect on the shrimp. If the harvest process is expected to take more than 1 h, harvested shrimps in hauling tanks should be kept on flow-through to reduce the metabolite concentration. Shrimp are transferred to the grow out pond when the biomass in the hauling tank is at maximum tank-carrying capacity. If the ambient air temperature is high (>25°C) and the grow out ponds are located far from the nursery facility, a decrease in water temperature is needed to reduce shrimp metabolism. A decrease in water temperatures of 20 to 22°C at a rate of 2 to 3°C every 15 min is adequate. Water temperature for a trip of up to 4 h can be decreased only to 23 to 24°C. The dissolved oxygen level should be measured every 15 to 30 min to ensure over saturation.

(xv) Troubleshooting in Raceways

The following is a list of some potential problems in an intensive raceway based nursery system.

1. Dead post-larvae are floating in the water column or on the bottom, the day after stocking.

   Possible explanations: (i) inadequate acclimation procedures, (ii) poor seed stock quality, (iii) poor shipping conditions, (iv) airlift pumps operating at too high a pumping rate, (v) bad water quality and (vi) disease outbreak.

   Suggested solution: If the acclimation process, airlift pumping rate, and water quality are adequate, the signs for external diseases need to be checked, if none are found, a representative sample should be taken and preserved in Davidson's fixative for histopathology. If mortality is estimated to be over 30 per cent, the animals may be discarded and restock with a new batch.

2. Two days after stocking, most post-larvae have empty guts.

   Possible explanations: (i) inadequate shipping or acclimation process, (ii) insufficient food, (iii) inadequate water quality and (iv) poor seed stock quality.

Suggested solution: If the acclimation records indicate an adequate process, there are enough *Artemia* nauplii in the raceway water, and water quality is adequate, the seed quality may not be adequate. If mortality exceeds 30 per cent in the following days shrimps need to be discarded and restocking is to be done.

3. A significant percent of the postlarvae have empty guts.

   Possible explanations: (i) insufficient food, (ii) inadequate water quality and (iii) disease outbreak.

   Suggested solutions: The raceway bottom should be checked for leftover feed. If no feed is found, the feeding records are checked to ensure that feed was provided on time; if positive, ration may be increased. If a lot of feed is untouched, water quality should be checked and compared with that of recent readings. Sudden changes in water colour, salinity, temperature, pH, dissolved oxygen and ammonia levels if any should be taken into consideration. If these parameters are adequate, check whether a new batch of dry feed had been used lately. If positive, switch back to the old batch and see if the feeding behaviour resumes. If none of the above measurements result in a satisfactory answer, check a few post-larvae under a dissecting microscope for disease signs. Once the suspected cause is identified, changes can be taken to correct the situation.

4. Many post-larvae are found dead on the raceway sidewalls.

   Possible explanations: (i) sudden light or loud noise interruption during the night, (ii) light was not turned on at night and (iii) low dissolved oxygen level.

   Suggested solutions: Identify the cause and correct the situation. If low dissolved oxygen is identified as the cause, inject oxygen during the night.

5. Post-larvae are jumping and/or swimming close to the water surface during the daytime.

   Possible explanations: (i) unfavourable water quality conditions, possibly with too high or low a dissolved oxygen level and (ii) disease outbreak.

   Suggested solutions: Check the previous and current water quality records, with emphasis on the dissolved oxygen level. If the water quality is adequate, look for disease signs.

6. Large numbers of post-larvae have broken legs and/or antennae.

   Possible explanation: Airlift pumps are working at too high a pumping rate.

   Suggested solution: Reduce the pumping rate so that no more than one or two post-larvae pass the airlift pump in a 30 sec. period.

7. The shrimp weight increment is lower than expected rate.

   Possible explanations: (i) sampling error, (ii) inadequate feed quality, (iii) inadequate water quality (including too high/low a temperature), and (iv) disease exhibition.

Suggested solutions: Verify that the sample drawn is representative of the population by taking another sample. Look closely at the size frequency distribution in the population, especially with large animals. If no sampling error is detected, check the feed shipment records (how long the current feed batch was in use and whether a new batch had been used). If negative, check the water quality and correct it if needed. Supplemental feeding with live/frozen adult *Artemia* is recommended to enhance growth. Check random samples for disease signs.

8. Post-larvae have heavy integument fouling (sessile ciliates, filamentous bacteria, benthic algae, etc.).

   Possible explanations: (i) poor growth performance and (ii) inadequate water quality.

   Suggested solutions: Enhance the molting rate by feeding with live/frozen adult *Artemia* and increase the water exchange. If molting does not occur within 2 to 3 d, a drastic drop in salinity for about 4 h can be used as a molt booster. Chemical treatment (such as a formalin bath of 200 ppm for 4 h) may be considered as a last resort.

9. Anoxic conditions develop on the raceway bottom.

   Possible explanations: (i) limited water circulation close to the raceway bottom and (ii) overfeeding.

   Suggested solutions: Increase the operation time of the bottom manifold, and if necessary, stir up the raceway bottom. Increase the water exchange rate and reduce the ration size.

10. A high amount of leftover feed is found on the raceway bottom.

    Possible explanations: (i) unfavourable water quality condition (especially low dissolved oxygen level), (ii) overfeeding and (iii) disease outbreak.

    Suggested solutions: Check the water quality records, skip one feeding, and increase the water to flush out part of the uneaten feed. Check a random sample for disease signs.

11. There are post-larvae in the sand filter chamber.

    Possible explanations: (i) inadequate mounting of the outlet filter pipe, (ii) rupture or hole in the filter pipe screen and (iii) use of inadequate filter pipe.

    Suggested solutions: Check with the underwater viewing apparatus for adequate mounting of the filter pipe. Pull the filter pipe out and check it for holes.

12. No water is coming out of the airlift pump.

    Possible explanations: (i) bent airline which is restricting airflow, (ii) airlift pipe is blocked by algae or other objects and (iii) malfunctioning valve.

    Suggested solutions: Check the hose, valve, and outlets to ensure free airflow.

13. The air diffuser does not produce a large amount of small air bubbles.

    Possible explanations: (i) air valve is closed, (ii) air pressure is too low and (iii) air diffuser is heavily fouled.

    Suggested solutions: Check the air valve and look for leaks in the distribution system. If heavy fouling is the problem, replace the air diffuser with a new unit.

14. The pH of the raceway water is above 9.7.

    Possible explanation: High algal densities.

    Suggested solutions: Decrease the light intensity and water temperature by shading the raceways. If the water temperature in the raceway intake fluctuates during the day, water exchange should be done during the cool hours.

15. The pH of the raceway water is below 7.

    Possible explanations: (i) high shrimp biomass and (ii) heavy bacterial load.

    Suggested solutions: Increase the water exchange rate and inoculate the raceway with algae from another raceway to promote algal bloom.

16. There is a bloom of blue-green algae in the raceway water.

    Possible explanations: (i) too high solar radiation, (ii) too high water temperature, and (iii) too low unicellular algal bloom.

    Suggested solutions: Decrease the light intensity and water temperature by shading the raceways. If the water temperature in the raceway intake fluctuates during the day, water exchange should be done during the cool hours. Scrape the raceway walls with a squeegee if necessary and flush heavily to remove the blue-green algae. Inoculate the raceway with water from an algal-rich source and fertilize with half of the initial recommended fertilizer dosage.

17. The water level in the raceway dropped by 10 cm overnight.

    Possible explanations: (i) leaking standpipe and (ii) hole in the raceway wall/bottom.

    Suggested solution: Identify the cause and prevent the leaking. In certain cases, only a temporary repair can be done until the harvest.

18. The oxygen injection system does not provide oxygen.

    Possible explanations: (i) empty O_2 cylinders/tanks, (ii) pump malfunctioning, (iii) oxygen leaking, and (iv) oxygen diffusers heavily fouled.

    Suggested solutions: If no leaking is detected and there is no oxygen shortage, check the pump. If the pump is OK, replace the O_2 probe diffuser.

19. Post-larvae have opaque tails.

    Possible explanations: (i) stress due to inadequate water quality and (ii) disease outbreak.

    Suggested solutions: Check the water quality; if adequate, check for disease.

Chapter 8
Diseases and Health Management

The careful planning and avoidance of diseases are crucial factors for successful shrimp culture operations. Shrimp can be infected by a large number of diseases caused by viruses, bacteria, fungi, protozoa, nutritional deficiency and abrupt changes in environmental parameters.

Shrimp health management is focused on disease prevention through proper nutrition, maintenance and stress reduction. Good management techniques include minimizing water quality fluctuations, maintaining stable micro algal culture etc. Post larval health should be evaluated on a daily basis. A random sample of 50-100 PL should be checked. The investigation should include information on PL deformities, abnormal swimming activity, hepatopancreas's colour and volume, gut content, fouling level, amount of fecal strands, moulting rate, animal stress signs and other unusual behaviour. Details of sampling should always be recorded.

Daily Observations

1. Plankton blooms, colour of the water, behaviour of shrimp.
2. Shrimp with empty guts, excess feed on trays.
3. Shrimp with soft shells or heads, fouling on shell.
4. Shrimp with deformities, discolouration of shrimp appendages, dense white muscles in shrimp.
5. Brown or black gills in shrimp.
6. Shrimp with bent or cramped tails.
7. Pond bottom condition.
8. Stressors like unionized ammonia and nitrate in the presence of high pH, insufficient oxygen and carbon dioxide levels, extreme changes in salinity, temperature, pH, alkalinity, molting, excess handling, presence of heavy metals and parasitism.

Pathogen Detection

For shrimp farmers, the most important aspect of disease diagnosis is the ability to detect the pathogen and minimize the losses. The diagnostic process involves detecting the agent responsible for the disease and its contribution to the disease. Rapid diagnostic tools like staining, haemolymph smears, squashes, bacterial cultures and gill and hepatopancreas exams can be used to monitor health and for diagnosis. Farms should use outside laboratories for analyzing molecular assays like dot blots, *in situ* hybridization and polymerase chain reaction (PCR) to confirm diseases.

Viral Diseases

There are over 14 types of virus diseases have been observed till recently in the cultured penaeid shrimps. These viruses are:

1. *Baculovirus penaei* (BP)
2. Monodon–type baculovirus (MBV)
3. Plebejus baculovirus (PBV)
4. Baculoviral mid-gut gland necrosis (BMN)
5. Type C baculovirus of *P. monodon* (TCBV)
6. Hemocyte infecting baculovirus of *P. monodon* and *P. esculentus* (HB)
7. Infectious hypodermal and hematopoietic necrosis virus (IHHNV)
8. Hepatopancreatic parvo like virus (HPV)
9. Lymphoidal parvo like virus (LOPV)
10. Lymphoid organ vacuolization virus (LOVV)
11. REO-3 and REO-4 (Reo virus)
12. White Spot Syndrome Disease (WSSV)
13. Taura Syndrome Virus (TSV)
14. Gill–Associated Virus (GAV)

There are no reports for successful curative measures for viral diseases. Hence utmost care needs be taken to prevent the contamination/spread of diseases into the culture systems.

1. Hepatopancreatic Parvo Like Virus (HPV)

HPV is observed in epizootics with high mortality rates. It has been found in dual infections with MBV in post-larval and juvenile *P. monodon* samples.

Gross Signs

It may not be specific. The severe infection exhibit the signs like whitish and atrophied hepatopancreas, poor growth rate, anorexia, reduced preening activity, gill fouling and secondary infections by Vibrio spp.

Transmission

HPV is transmitted either from parent brood stock or from shrimp to shrimp with efficiency only during the larval or early post larval stages.

Host Range

P. merguinensis, P. chinensis, P. penicillahis, P. esculenlus, P. monodon, P. semisulcatus

2. Infectious Hypodermal and Haematopoietic Necrosis Virus (IHHNV)

It is the smallest of the known penaeid shrimp viruses, would cause serious disease. It is an acute disease causing agent and effecting very high mortalities in juveniles. Vertically infected larvae and early PL do not become diseased. In horizontally infected juveniles, the young juveniles being the most severely affected; infected adults show signs of disease or mortalities.

Causative Agent

It is caused by a non enveloped icosahedral virus, Family: Parvoviridae

Host Range

L. vannamei, P. stylirostris, P. occidentalis, P. monodon, P. semisulcatus, P. japonicus, P. californiensis, F. indicus and *P. merguiensis.*

Gross Signs

Reduction in food consumption, rising slowly to water surface. White or coloured spots on the cuticular epidermis especially at the junction of the tergal plates of the abdomen, giving such shrimp a molted appearance; Runt deformity syndrome (RDS) has been linked to IHHN ; Juvenile shrimp with RDS show reduced growth rates, bent or deformed rostrome, wrinkled auleunal flagella and anticular defrormities.

Confirmatory Tests

Molecular methods like Dot blot hybridization and PCR are used for non-lethal screening of non clinical brood stock and juveniles of susceptible species.

Gross Observation

Gross signs are not IHHN specific. Shrimp may rise slowly to the water surface, become motionless and then roll over and slowly sink to the bottom, white or buff coloured spots (which differ from WSSV) in the cuticular epidermis especially at the junction of the abdominal tergal plates, resulting in a mottled appearance.

3. Lymphoidal Parvo Like Virus (LOPV)

Affected shrimp had multinucleated giant cells in their hyper trophied lymphoid organs. Giant cells showed nuclear hypertrophy, marginated chromatin and formed fibrocyte encapsulated spherical structures. Electron microscopic studies revealed the presence of 25-30 mm diameter virus like particles.

Host Range

P. monodon, P. merguiensis, P. esculentus.

4. The REO Like Viruses

Hepatopancreas is the principal target for these REO 3 and REO 4 viruses. It has been found in shrimp with mixed infections by other viruses or fungi or disease syndromes.

Host Range

P. japonicus, P. monodon, P. chinensis, L. vannamei.

5. The Toga Viruses: Lymphoid Organ Vacuolization Virus (LOVV)

The lymphoid organ is the target for the Toga viruses.

Host Range

P. monodon, P. esculenhis, L. vannamei, P. stylisoslsis, P. penicillatus, P. chinensis.

6. *Baculovirus penaei* (BP-type Baculoviruses)

They were enteric infecting the epithelial cells of the hepatopancreas and mid gut of their host results in high mortality rates. Infections from parent to offspring apparently results from faecal contamination of spawned eggs by virus contaminated faeces. Horizontal transmission result from faecal contamination or from cannibalism. It causes significant disease problems in the larval, post larval and early juvenile stages of several penaeid species.

Host Range

L. vannamei, P. penicilahus, P. stylirostris.

7. Monodon–Type Baculo Viruses

MBV was named after the penaeid species in which it was just recognized. A similar agent found in *P. plebejus* and thus called *Plebejus baculovirus* (PBV).

MBV is not a highly virulent pathogen of *P. monodon.* It is often found in apparently healthy prawns even at relatively high severity indices of infections. However their presence along with WSSV could cause serious problems to the aquaculturists.

Control Measures

1. Overcrowding, chemical and environmentally induced stress will increase the virulence of MBV and BP infections.
2. Disinfection of naupli with Formalin (400 ppm; 1 min); Iodopohore (0.1 ppm; 1 min) and filtered clean sea water (3-5 min).

8. BMN (Baculoviral Midgut Gland Necrosis) and Other Type C Baculoviruses

BMN disease in larval stages or penaeids is characterized by a sudden outset and a high mortality rate. A "white turbid" hepatopancreas (midgut gland) in larvae and PL is the first grossly visible sign of the disease. Severely infected PL are easily distinguished because they float inactively on the surface of the water and exhibit a white midgut line through the abdomen.

Causative Agent

Baculoviral midgut gland necrosis virus (BMNV).

Host Range

BMN was observed as natural infections in *P. japonicus, P.monodon, P. plebejus* and as experimental infections in *P. chinensis* and *P. semisulcatus.*

Control Measures

The concentrations of various disinfectants required to kill BMNV are toxic to shrimp larvae. Complete or partial eradication of viral infection may be accomplished through washing of naupli using clean sea water to remove the adhering excreta.

9. White Spot Syndrome Virus (WSSV)

Causative Agent

It is called as White spot syndrome virus (WSSV) or white spot virus (WSV), a double stranded DNA (dsDNA) virus. WSSV was described as a non-occluded baculovirus but subsequent analysis of WSSV-DNA sequence does not support this contention. The viruses in this complex have recently been shown to comprise a new group with the proposed name of nimaviridae (Van Hulten *et al.*, 2001)

Host

It has a wide spectrum of hosts. Out breaks were first reported from farmed *P. japonicus* in Japan, *P. chinensis, F. indicus, P.merguiensis, P. monodon, P. setiferus, P. stylirostris* and *P. vannamei.*

Signs

Acutely affected shrimp demonstrate anonexia and lethargy, loose cuticle with numerous white spots (about 0.5–2 mm in diameter) on the inner surface of the carapace. These spots are within the cuticle structure and cannot be removed by scraping. Moribund shrimp may also show the pink to red colouration pathology which is associated with systemic destruction of the ectodermal and mesodermal tissues of the gills and subcuticular tissues.

Comfirmatory Tests

An accurate diagnosis can be accomplished by the nested PCR of tissues and haemolymph, insitu hybridization and by immuno diagnostic techniques.

Gross Observation

Cessation of feeding followed by swimming near surface of the water. White inclusions are embedded in the cuticle and often show reddish discolouraiton of the body. The cuticular inclusions range from minute spots to a size of several mm in diameter that may coalesce in to larger plagues.

Control Measures

There are no known effective treatments available, however a number of preventive measures are recommended to reduce the spread of this disease.

1. All infected individuals and their offsprings should be destroyed systematically.
2. PL should be screened by Nested PCR using sufficiently large numbers of PL to ensure detection of significant infections.
3. Rapid changes in water temperature, pH, hardness and salinity or reduced oxygen levels (< 2 ppm) for extended periods can trigger out breaks of WSSV in shrimp.
4. Affected raceways should be treated immediately with 30 ppm chlorine to kill the infected shrimp and any potential carriers.

10. Yellow Head Disease (YHD)

Causative Agent

YHD is caused by the yellow head virus (also reported in older literature as Yellow Head Baculo virus (YBV) and Yellow Head Disease Baculo Virus (YHDBV). It is row known not to be a member of the Baculoviridae, Family : Coronaviridae.

Host Range

Natural infections occur in *P. monodon* but experimental infections were possible in *P. japonicus, L. vannamei, P. seiferus, P. aztecus, P. duoradum* and *P. stylerostris.*

Signs

Diseased shrimp aggregate in the edges of the ponds or near the surface. The hepatopancreas becomes discoloured which gives the cephalothorax a yellowish appearance, hence the name of the disease. The overall appearance of the shrimp is abnormally pale. Gross signs and mortality occur within 2 to 4 days following an interval of exceptionally high feeding activity that ends in abrupt cessation of feeding. Mortalitiies can reach 100 per cent within 3–5 days. Clinical signs are not always present and their absence does not rule out the possibility and YHD injection.

The Oka organ, gill, heart and subcutical tissues including those of the stomach epithelium, contain the highest level of YHV.

Confirmatory Tests

Reverse Transcriptase PCR (RT-PCR) technology is recommended for certification of YHV infectious strain of broodstock and fry.

Gross Observation

YHD can be suspected when the shrimp show an abnormal increase in feeding rates, slow swimming behaviour, pale overall body colouraiton, a yellowish cephalothorax, pale gills and hepatopancreas.

Control

1. Screening for YHD for incoming seeds

2. Infected individuals and their offspring be destroyed in a sanitary manner.
3. Avoidance of rapid changes in pH or prolonged periods of low dissolved oxygen, water pH levels less than 9 should be avoided.
4. Avoid fresh aquatic feeds unless the feed is subject to prior sterilization (gamma radiation) or pasteurization.

11. Taura Syndrome (TSV)

Causative Agent

Taura Syndrome Virus, member of Picornaviridae.

Host Range

The most susceptible species are *L. vannamei, P. stylirostris* and *P. setiferus*. Post larvae and juvenile of *P. monodon, P. chinensis, P. duorarum* have been infected experimentally.

Signs

Post-larvae or older shrimp may show a pale reddish colouration especially at the tail fan and pleopods. Shrimp showing signs have soft shells, an empty gut and often die during moulting.

Confirmatory Tests

Histopathology, Transmission Electron Microscopy (TEM), Dot Blot, *in situ* Hybridization and PCR probes are used as confirmatory tests.

Control

1. Wild caught post-larvae have increased tolerance to TSV than hatchery reared PL.

2. Selective breeding.
3. Eradication depends on total removal of infected stocks disinfection of the culture facility and avoidance of reintroduction of virus.

12. Gill Associated Virus (GAV)

Causative Agent

It is a single stranded RNA virus related to viruses of the family Coronaviridae. GAV can occur in healthy or diseased shrimp and was previously called Lymphoid organ virus (LOV).

Host Range

Natural infection is seen in *P. monodon* and experimental infection in *P. esculentus, P. merguiensis* and *P. japonicus.*

Gross Observation

Shrimp with acute infection are lethargic, lack of appetite and swim on surface or around the edges of the ponds. The body may develop a dark red colour particularly on appendages, tail fan, mouth parts and gills tend to be yellow to pink in colour.

Confirmatory Tests

Reverse Transcriptase Polymerase Chain Reaction (RT-PCR), Histopathology and Transmission Electron Microscopy (TEM).

Bacterial Diseases

Rickettsial and Chlamydial Diseases

Reports of Rickettsia or Rickettsial–like bacteria in penaeid shrimp suggest that these organisms are important pathogens of penaeid shrimp in wild and cultured shrimps.

Reports on chlamydial–like agent observed in the cytoplasm of hepatopancreatic cells of *P. japonicus* which were infected by BMN are available.

Gross Signs

Lethargic, inappetence, poor escape responses and white coloured hepatopancreas. The disease called as Texas pond mortality syndrome (TPMS) was reported from Texas due to Rickettsia–like agent in *P. vannamei* where 50 per cent to 99 per cent stock were affected.

Vibriosis

Vibrio spp were found to constitute the majority of the bacterial pathogen associated with the gut, gills or cuticle of wild or cultured penaeid shrimp. This opportunistic Vibrio spp. establish lethal infections as a result of infectious diseases, nutritional diseases, extreme environmental stress, wounds etc. Bacterial isolates from diseased penaeid shrimp include species, such as *V. alginolyticus, V. parahaemolyticus, V. harveyl, V. splendidus, V. vulnificus* and *V. damsella.*

Bacterial Shell Diseases (BSD)

BSD lesions are typically brownish to blackish in colour causing, single or multiple eroded areas on the general body cuticle, appendages and gills. The black pigment in shell disease lesions is due to melanin, which is an end product of the crustacean inflammatory response. If such lesions are not successfully resolved by the affected shrimp's inflammatory response, septicemia and death will result. The disease is called as brown spot shell disease, burned spot disease, rust disease, shell disease or black spot disease. *Vibrio, Pseudomonas* and *Beneckea* have been known to cause this disease.

Control Measures

Provision of better water quality, reduce the organic load in the system by increased water exchange and Feeding terramycin incorporated feed at 0.45 mg/kg feed.

Bacterial White Spot Syndrome (BWSS)

Causative Agent

Bacillus subtilis is the possible causative agent. Bacterial white spots are less dense than virus induced white spots.

Host Range

The syndrome has only been reported in cultured *P. monodon.*

Sign

Dull, rounded white spots are seen on the carapace and all over the body. Wet mount microscopy reveals the spot as opaque brownish lichen like lesions, spot centre is often eroded and even perforated.

Diagnostic Methods

Negative WSSV–PCR results from samples showing gross clinical signs attributed to WSSV, may be suggestive by the alternate aetiology of BWSS.

Control

1. Build up of high bacterial density in rearing water, indiscriminate use of probiotics containing *Bacillus subtilis* shoud be avoided.
2. BWSS in shrimp pond can be treated with quick lime (CaO) at 25 ppm and the use of CaO may itself cause problems due to rapid changes it can cause in water pH.

Mycobacterium Infections

They exhibit abnormally black pigmentation in areas of the body or melanized granulomatous lesions in lymphoid organ, cuticle, heart, hepatopancreas, antennal gland, ovary and gills. No control methods or therapy for mycobacterium infection in penaeid shrimp have been devised.

Fungal Diseases

Larval Mycosis

The fungi that cause this group of diseases are all phycomycetes. Two genera of phycomycetes, Lagenidium and Sirolpidium, are the best known of this group of shrimp pathogens. These fungi grow unrestricted with in infected larvae, and in a matter of only a few hours after initiation, the infection is well established. The encysted spore germinates and mycelium grows replacing the host's muscles and other soft tissues. Two chemotherapeutics have shown efficacy in controlling larval mycoses: malachite green oxylate 6 to 10 ppb) and trifluralin (10 to 100 ppb). Calcium hypochlorite at concentrations below 100 ppm was found to be ineffective against Lagenidium spp and 500 ppm for 24 hour was found to be the mycocidal dose.

Fusariosis

Infections due to *F. solani, F. moniliforme* and *Fusarium* spp infections have been reported from decapod crustaceans. Except for a single report of infection of *P. japonicus* by *F. moniliforme,* all other reports of *fusariosis* in cultured penaeids have been by *F. solani.*

Host Range

P. monodon, P. japonicus, P.californiensis, P. setiferus, P. stylirostris and P. vannamei.

Diagnosis

It can be accomplished by direct microscopic demonstration of the canoe shaped macroconidia which are characteristic for *Fusarium* spp.

Treatment

No effective chemotherapy or preventive measures are known for Fusariosis disease. More recently a commercially available microbiocide (Rohm and Haas' compound BG 101â, which contained the biocides 5-chloro-2-methyl-4-isothiozolin-3-one) showed some efficacy in laboratory experiments.

Parastitic Diseases

Microsporidians

It cause a group of diseases in penaeids that are collectively called "cotton" or "milk shrimp disease". *Ameson nelsoni, Agmasoma penaei, Thelohania duorara* and *Pleistophora* sp are the microsporidans affect the penaeids in different geographic regions. A conditioning intermediate host is needed for transmission. These microsporidrans infect striated muscle, gonads, adjacent connective tissues and occasionally blood vessels, gills and the hepatopancreas. Infection of the gonads cause white hypertrophied gonadal organs, but infection of other tissues, such as the subcutis can result in tumour–like swellings of the overlaying cuticle. The infected shrimps have distinctly opaque musculature ovaries and often develop dark blue or blackish discolouration due to the expansion of the cuticular chromatophores.

Treatment

Oral administration of buquinolate.

Haplosporidians

The first reported haplosporidian infection was in the hepatopancreas of juvenile of *P.vannamei.* Infected shrimps showed poor growth and histopathology of infected hepatopancreas showed the parasite to be present in the cytoplasm of tubule epithelial cells in which moderate to heavy hemocytic inflammation had occurred. Species generally affected are *L. vannamei* and *P. monodon.*

Gregarines

Gregarines (Protozoa, Apicomplexa) are common parasites. Gregarines may occur inter-or intracellularly and most species are not considered to be highly pathogenic. *Nematopsis* spp and *cephalolobus* spp commonly occur in the penaeid shrimp. In shrimp, ingestion of an infected intermediate host, which contains spores of the gregarine results in infection. Medicated feeds which contain anticoccidal drugs at dose rates used in poultry or cattle are now routinely used by shrimp farmers in Ecuador.

Disease Caused by Epicommensals

Leucothrix mucor, Zoothamnium spp., Epistylis spp are the common epicommensal organisms encountered in culture shrimps.

Filamentous Fouling and Gill Diseases

L. mucor and *Thiothrix spp.* readily attach to the body surfaces of shrimp. These organisms attach favourably to the gills and accessory gill structures in juvenile or older penaeid shrimp.

Non-infectious Diseases

Ascorbic Acid Deficiency Syndrome (Black Death Disease)

The typical large black (melanized) lesions that occur in shrimp due to a dietary deficiency of ascorbic acid is called as the ascorbic acid deficiency syndrome (AADS). Shrimp with AADS typically display blackened (melanized hemocytic) lesions in tissues with a high content of collagen. Black death disease has not been observed in sub-adult or adult shrimp and it is apparently confined to the juvenile stages of the species.

Cramped Muscle Syndrome

CMS of penaeid shrimp has also been called "Cramped tail", body cramp and bent tail. CMS is characterized by an autemortem ventral flexure of the abdomen which is so rigid that it cannot be straightened without tearing the abdominal muscle tissue. Severe CMS typically results in the rapid death of affected shrimp. In shrimp with a partial abdominal flexure, focal areas of white muscle developed in the abdomen.

Chronic Soft Shell Syndrome

They are typically dark coloured and heavily colonized by surface fouling epibionts, such as the colonial peritrich *Zoothamnium* sp and other epicommensals.

Soft-shelled shrimp are lethargic, weak and susceptible to wounding and cannibalism, show poor growth rates and eventually die. Improper storage of feeds, use of rancid or low quality feeds and inadequate feeding rates were noted to be closely associated with chronic soft shell syndrome.

Pale Coloration or Blue Disease

It is also called as pale colouration, blue disease, sky blue shrimp disease and blue shell syndrome. It is due to adverse soil and water quality conditions. Penaeid shrimps such as *P. stylirostris* and *P. monodon* are generally affected by this problem. The affected shrimp show pale or a little or no brown (in *P. monodon)* or grey brown (*P. stylirostris)* pigmentation. This may be prevented or corrected by the use of feeds with a high content of carotenoids and vitamin A.

Toxic Disease Syndromes

Aflatoxicosis

Aflatoxin cause inflamed lesions in the hepatopancreas of penaeid shrimp (*P. stylirostris* and *L. vannamei)*. Hepatopancreatic lesions are almost identical to those observed in aflatoxicosis occur in *P. monodon* with a disease syndrome called "Septic Hepatopancreatic Necrosis" (SHPN).

Black Gill Disease

A number of disease syndromes of cultured penaeids are accompanied by the presence of black (melanized) inflamed lesions in the gills. Besides fungi, certain protozoans and the AADS, certain heavy metals, oil and other chemical irritants can also cause the syndrome. Black gills or melanised gills (burned gills, "black spot disease" etc.) and related melanized criticular lesions such as shell disease in shrimp are common manifestations of a number of disease syndromes which are regularly observed in wild and cultured marine and fresh water shrimp. The black or brown pigment present on the gills, appendage and cuticular lesions is melanin, which is formed at the sites of hemocytic inflammation and tissue necrosis. Copper based algicides are frequently used in shrimp culture and their misuse may result in necrosis and melanization of the gills and in cuticular lesions. The most common causes of black gills and occasional cuticular lesions in cultured shrimp are ammonia and nitrite.

Gas Bubble Disease

It occurs due to the super saturation of atmospheric gases and oxygen. Gas bubble disease in shrimp due to atmospheric gas (primarily nitrogen, b/c air is 80 per cent nitrogen) has been linked to faulty plumbing and pumps.

Oxygen saturation cause gas bubble disease in penaeid shrimp. It occurs when dissolved oxygen level reaches or exceeds 250 per cent normal saturation in sea water of 24 to 26°C and 35 ppt salinity.

Signs

Rapid, erratic swimming behaviour, stuporus behaviour, shrimp so affected tend to float near the water surface with the ventral side of the cephalothorax.

Muscle Necrosis

Spontaneous necrosis or idiopathic muscle necrosis is characterized by whitish opaque areas in the striated musculature, especially of the distal abdominal segments. The condition typically follows periods of severe stress from over crowding, low dissolved oxygen levels, sudden temperature or salinity changes, rough handling etc. When irreversible damage occurs and opportunistic bacteria invade the necrotic areas of the abdomen, the condition is called "tail rot".

Chapter 9

Biotechnological Applications in Shrimp Aquaculture Management

Biotechnology is a general term that refers to any endeavor using applied biology as the basis of technology. The impact of biotechnology on the aquaculture management has been significant. Biotechnological innovations are everyday occurrences and there is a significant economic incentive to bring applicable technologies into aquaculture.

1. Water Quality Management

(a) Bioremediation

Bioremediation is a biotechnological process of using selected micro or macro-organisms to reduce harmful waste to less hazardous levels. *Bacillus* sp. *Vbiro* sp., *Nitrosomonas* sp., *Nitrobacter* sp., *Aerabacter aerogenosa, Cellulomonas biozotca, Saccharomyces* sp.and several micro algae are some of the biological agents recognized for their bioremediation abilities.

Naturally occurring micro-flora might not be able to degrade the accumulated wastes efficiently in intensive aquaculture system. Therefore supplementing them with selected bacteria in sufficient numbers might be useful. Inoculated bacteria initially act on the organic matter and produce ammonia that is subsequently oxidized to nitrite and nitrate. Introducing active aerobic bacterial populations could be an effective strategy in rapid degradation of complex organic compounds and overall improvement of health status of cultured organisms. Therefore, it is necessary to formulate the product with selected micro-organisms for bioremediation.

Selection of Suitable Bacteria

Suitable species of bacteria need to be identified and isolated for use as bioremediators. The characteristic features include :

1. It should be non-pathogenic in nature
2. It should have fast multiplication rate
3. It should not be a fastidious organism
4. It should not alter the beneficial microflora of the aquatic system.

(b) Biofilters

In almost all aquaculture systems there is some degree of biological filtration taking place , planned or not. Some systems, especially closed salt water systems, rely on biofilters almost exclusively. Biofilters primarily designed to remove dissolved nutrients from the water. Most importantly they convert the harmful nitrogenous forms such as ammonia to less toxic form like nitrate.

The primary biological agents in these filters are autotrophic bacteria, although algae, yeast, protozoan and some tiny animals may also be present and aid in the removal of the wastes from the water. These autotrophic bacteria are not found on the surface of filter particles as individuals. They form colonies and secrete a coating of slime for protection.

Types of Biological Filter

Submerged Bed or Under Gravel Filter

Gravel and crushed coral are often used as medium for the growth of nitrifiers and the added advantage is that, they have the capacity of buffering the pH since they are made up of calcium carbonate. As the water passes through the bed, the bacteria take up the waste and oxidized into nitrate. The major drawback of this type is that the reaction is oxygen limited and if there is not enough oxygen, the conversion will be slow or not taking place at all.

Trickling Filters and Rotating Drums

Trickling filters are never oxygen limited and therefore often more efficient than submerged filters. In rotating drum, only a part of the filter will be submerged and the other part is passing through the air. Drums are covered with a sort of mesh covering and are usually filled with light weight plastic or polypropylene materials designed to have a large growing area for microbes.

(c) Constructed Wetlands

Disposal of shrimp farm effluent is of great concern

now in India and efforts are being taken to tackle the technical and the managerial problems involved in this. Further the used water from an aquaculture system is a resource and it should not be wasted, instead it needs to be recycled.

What are Constructed Wetlands (CWs)?

Wetlands are the most important ecosystem on earth, due to their unique hydrological conditions and their role as ecotones between terrestrial and aquatic system. They are considered as the "kidney of landscape", since they have inherent capability to transform and store organic matter and nutrients. Because of this property, the wetlands precisely exploited for water quality treatment/ improvement.

Removal Mechanism

Sl.No.	Constituent	Removal mechanism
1.	Suspended solids	Sedmentation/filteration and degradation
2.	Nitrogen	Ammonification followed by microbial nitrification and denitrification
3.	Phosphorus	Soil sorption (adsorption-precipitation reactions with Al, Fe, Ca and Clay particles)
4.	Pathogens	Natural die-off, excretion of antibiotics from roots of macrophytes
5.	BOD	Microbial degradation and sedimentation

2. Water Treatment

(a) Primary Treatment

(i) Screening and Sedimentation

Screening is the most commonly used treatment to remove larger solids. Static screens and rotary drum screens are often employed. Sedimentation is used to remove suspended solids present in the effluent. Sedimentation or settling is based on the difference in density between the bulk of the liquid and solid particles. Discrete settling, flocculent settling or zone settling can achieve sedimentation.

(ii) Separation of Oil and Grease and Flotation

This can be accomplished by gravity separation provided the oily particles are large enough to float and are not emulsified.

Flotation removes both oily material and suspended particles. The most common procedure is that of dissolved air flotation. Flotation efficiency can be enhanced by the use of coagulating aids, in the form of chemicals.

(b) Biological Treatment

The objective of biological wastewater treatment is to remove non-settling solids and dissolved organic matter from the effluent by using microbial populations. These microbes are responsible for degradation of organic matter and stabilizing of the waste.

Aerobic Process

The most common aerobic processes are activated sludge systems, aerated lagoons, trickling filters etc.

Activated Sludge System

Activated sludge system involves aerobic degradation. In these systems, the organic load enters a reactor where the active microbial population is present and is continuously aerated. This mixture passes to a secondary settling tank where the cell mass is settled. The settled cell mass is often recycled.

Aerated Lagoons

These lagoons are basins normally excavated in the earth and operated without recycling of solids in the system. Accumulated solids, which settle on the bottom must be removed periodically.

Anaerobic Treatment

The anaerobic treatment of wastewater proceeds with degradation of the organic load to gaseous product viz. methane and carbon dioxide. The organic load present in the water is first converted into its soluble forms, which is consumed by acid producing bacteria to give volatile fatty acids. In this treatment the digesting tank must have good sealing to avoid entry of oxygen.

Physico-chemical Treatments

(i) Coagulation–Flocculation

In coagulation operation, a chemical substance is added to an organic colloidal suspension to cause destabilization by the reduction of forces that keep them apart. It also involves the reduction of surface charges responsible for particle repulsion. This charge reduction causes flocculation. Particles of larger size are then settled and effluent is obtained.

Disinfections

Chlorination

Chlorination is commonly used to destroy bacteria and algae in the effluent. During chlorination process, formation of chloramines poses a problem in that the waste water containing appreciable amounts of ammonia increases the amount of chlorine needed to achieve the desired degree of disinfections. Additional amount of chlorine result in the disappearance of chloramines by oxidation.

Ozonation

Ozone, a strong oxidation agent, has been used for disinfection due to its bactericidal property and also its potential to remove the viruses. Advantage of ozone over chlorination is that it does it produce dissolved solids and is not affected by ammonia compounds.

3. Feed Management

Microencapsulated Diet

Microencapsulated diets are specially prepared diets, which are comparable to plankton in size and performance. In a microencapsulated diet, the liquid and particulate dietary components are enclosed with a carefully engineered wall. The principle involved in the micro encapsulation is the release of the internal nutrients at active sites, within the target biological system. A variety of mechanisms are adopted to effect this process by the digestion of wall components. The wall can be digested by enzymatic action, pH alteration or bacterial action. The wall should be made of polymer like modified gelatin, so that the nutrients within the capsule could be released by enzymatic processes of the animals or by microflora present in the gut.

X–linked Protein Microencapsulate

This is prepared by emulsifying dietary components in aqueous solution of cyclohexane containing natural surfactant. Active sites of protein molecules within the diet are then cross-linked by interfacial polymerization using an acid chloride. The reaction is terminated and the cross-linked protein microcapsule is freeze dried. This process is being commercially employed by Frippack feeds of UK and a variety of capsules being produced ranging from CAR (Crustacean algal replacement) at 5.20 mm to CD (Crustacean diet) at 90–150 mm diameter.

Bioenrichment Diets

It is a process by which micronutrients or chemotherapeutants are incorporated into live feed organisms either directly or indirectly to ensure the complete acceptability of the incorporated nutrients/chemotherapeutants. Bio-enrichment used to incorporate nutrients like vitamins, essential fatty acids, minerals and amino acids, besides therapeutants like antibiotics. Bioenrichment can be carried out by two methods.

(i) Direct Methods

This method involves homogenizing the nutrients or chemotherapeutants with water and raw egg yolk and feeding the resultant emulsion to the live feed or organisms which act as carriers. The live feed incorporated with nutrient/therapeutants, is then fed to the animals.

(ii) Indirect Method

Baker's yeast contains a fairly high amount of monethylenic fatty acids and number of M-3 highly unsaturated fatty acids (HUFAs). A new type of yeast has been developed as a culture organism for rotifers in order to improve upon the nutritional value of the rotifers. This new type of yeast-designated u-yeast is grown in a culture medium enriched with essential fatty acids prior to being fed into the rotifers.

Incorporation of Yeast in Aqua Feed

Yeasts are single celled heterotrophic organisms and

the cell wall is made of beta 1.3 and beta 1.6 glucan, mannan protein and chitin. Yeast has been used as a vitamin supplement and as a raw material in animal feeds due to its high protein content about 35 to 45 per cent and high levels of B vitamins, Niacin and folic acid.

The two main types of yeast used in aquaculture feeds are brewers yeast and molasses yeast. Brewers yeast can be quite bitter and it needs to be debittered by mixing the yeast slurry with sodium hydroxide and sodium phosphate at pH 10 and 45–50°C.

Mannan Oligosaccharides

There has been some interest in the use of mannan oligosaccharide (MOS) in animal feeds as an alternative to antibiotics. MOS can bind with gram negative pathogens such as the enterobacteriaceae that have mannose specific Type I fimbriae and prevent them attaching to the cells of the intestine, thereby preventing colonization and infection. In addition, MOS can also act as an immune system.

Beta-Glucans

Beta-glucans are better known for their stimulatory effect on the immune system. Shrimp or fish given with a diet containing beta-glucans show better performance following bacterial challenges than those fed normal diets. The source and the processing methods of the beta-glucan are also important since activity is better when the integrity of the beta-1,3 glucan is retained.

Nucleotides

Yeast extracts provide a rich source of vitamins, which are important for many physiological functions including metabolism of amino-acids, carbohydrates and lipids. In addition, they are also a rich source of nucleotides. The dietary sources of nucleotides can have beneficial effects and may optimize the functions of rapidly dividing tissues, such as those of the immune system, which lack the capacity to synthesis nucleotides. Besides, it is essential during periods of rapid growth or cell replication such as early development of larvae.

4. Hatchery Techniques

Eyestalk Ablation

The process of unilateral eyestalk ablation is used commonly in marine shrimp hatcheries for maturation/ reproduction. This methods of inducing females to develop mature ovaries is used for two reasons: (*i*) most captive conditions cause inhibitions in females which keep them from developing mature ovaries in captivity and (*ii*) even in conditions where a given species will develop ovaries and spawn in captivity use of eyestalk ablation increases total egg production and increases the percentage of female population.

The most commonly accepted theory in eyestalk ablation is that a gonad inhibitory hormone (GIH) is produced apparently during non-breeding season and is absent or present only in low levels during the breeding season. The reluctance of most penaeids to routinely develop mature ovaries in captivity is a function of elevated levels of GIH, and eye stalk ablation lowers the GIH enabling the females to breed effectively.

Techniques of Eyestalk ablation

1. *Simple pinching*: Usually performed half to two thirds down the eyestalk. This methods may leave an open wound.

2. *Enuleation*: It is performed by slitting one-half to two-thirds down the eyestalk and moving distally until the contents of the eye have been removed.
3. Cauterizing through the eyestalk with either an electrocautery device or an instrument such as red-hot wire or forceps.
4. Ligation by tieing off the eyestalk tightly with surgical or other thread.

The following observations have been made concerning the use of eyestalk ablation in captive reproduction.

1. Captive spawn size is smaller than in wild-matured females.
2. Eyestalk ablation increases total egg production in captivity by producing more frequent spawning.
3. Reduced fecundity with successive captive generation.
4. Survival from nauplius to post larvae is sometimes lower in captivity–produced larvae Vs wild-produced.

5. Immunostimulants

Intensification of any culture system would lead to poor physiological environment for shrimp that could increase the susceptibility to infections. Various chemotherapeutants have been used to treat bacterial infections in cultured shrimp for the last 2 decades. However, the incidence of drug-resistant bacteria has become a major problem in fish culture. Vaccination is a useful prophylaxis for infectious disease of fish but against intracellular pathogens such as *Renibacterium salmoniarum* has not so far been successful.

Immunostimulant increases resistance to infectious diseases, not by enhancing specific immune response, but by enhancing non-specific defense mechanisms. Therefore, as there is no memory component involved, the response is likely to be of short duration. Use of these immunostimulants is an effective means of increasing the immunocapacity and disease resistance of shrimp.

Defense Enhancement by Immunostimulants

Shrimp treated with imunostimulants usually show enhanced phagocytic cell activities. The activity can be detected by phagocytosis, killing and chemotaxis. Lymphocytes are also activated by immunostimulants. Immunostimulants enhance the mitogen activities induced by concavalain A or lipopolysaccharides and produce macrophage activating factors. Complement activity can also be activated by several immunostimulants.

Immunostimulants Commonly Used in Shrimp

1. *Synthetic chemicals*: Levamisole, FK–565 and MDP (Muramyl dipeptide)
2. *Bacterial Derivatives*: β-glucan, peptidoglucan, FCA (Freund Complete Adjuvant) and LPS.
3. *Polysaccharides*: Chitin, Chitosan, Lentinan and Oligosaccharides.
4. *Nutritional factors*: Vitamin C and Vitamin E.

6. Probiotics

The use of the antibiotics as growth promoters is resulting in the development of resistant populations of bacterial and in turn restricting the use of antibiotics for therapeutic purposes. Therefore, the use of antibiotics as supplements in animal feeds is restricted and only permitted antibiotics and other chemical supplements should be used in feeds.

Probiotics are the opposite terms for antibiotics. They defined as live microbial feed supplement, which beneficially affects the host by improving its intestinal microbial composition and health of the host. The first probiotic organism discovered long time ago was *Lactobacillus* sp. the lactic acid producing bacteria. They were thought to prevent colonization of the gut by other disease causing bacteria by a process known as competitive exclusion. The most commonly used microbes as probiotics are *Lactobacillus acidophilus, L. delbreuck, L. casei, L. reuteri, Streptococcus faecinum, S. salivarus, S. lactis, Pediococcus* selected strains of *Bacillus* sp. and certain strains of yeast belonging to *Saccharomyces* sp.

Probiotics show antagonism to other organisms through (*i*) competition with other species for binding sites: (*ii*) specific antagonism against other species some of the antagonistic bacteria against fish and shellfish pathogens include *Bacillus* sp., *Roseobacter* sp. *Carnobacterium divergens,* against pathogens like *Vibrio harvei; V.anguillarum, Edwardasiella, Aeromonas* and *Pastrella.*

7. Disease Management

PCR

The Polymerase Chain Reaction (PCR) is an *in vitro* technique which allows the amplification of specific DNA that lies between two regions of known DNA sequences. The selective amplification is achieved by using oligonucleotide primers, also known as amplimers. These are short, single stranded DNA molecules which are complementary to defined sequence of DNA template. Strand synthesis is brought about by a thermostable DNA polymerase enzyme, which adds de-oxyribonucleotide triphosphate (dNTPs), successively to the 3′OH end of pimers copying of a specific genetic target is accomplished in three step reaction.

1. Denaturing the target DNA at elevated temperature (90–95°C).
2. Cooling the reaction to promote annealing of oligonucleotide primers that are complementary to either strand of target DNA.
3. Extending the bound primers by DNA, polymerase action resulting in the replication of target sequences.

The reaction is repeated in successive cycles of heating and cooling referred to as thermo cycling. After a series of cycles an exponential amplification can be achieved. The final PCR product can be characterized in many ways. The most common being to determine the size of amplified fragments using gel electrophoresis.

Presently, PCR methods are available for the detection of many pathogens of shrimp viz. *V. vulnificus, V. parahaemolyticus, V. penaeida, WSSV,* MBV, IHHNV, TSV, rod shaped nuclear virus of *P. japonicus* and BP. The typical amplification reaction includes:

1. The sample of target DNA
2. Two oligonucleotide primers
3. Deoxynucleotide triphosphates (dNTPs)
4. Reaction buffer

5. Magnesium and optical additives
6. Taq DNA polymerase
7. Double distilled water.

The components of the reaction are mixed and placed in an automated instrument. The thermocycler that takes the reaction to the series of different temperatures for varying amounts of time. This series of temperature and time is referred to as one cycle amplification. In each cycle of amplification, the quantity of the target DNA would be doubling and after 20 cycles it would generate approximately a million copies of DNA.

Monoclonal Antibody

Antibodies produced by hybridomas are known as monoclonal antibodies. Monoclonal antibodies are highly specific and can be produced in unlimited quantities. The usefulness of monoclonal antibodies stem from three characteristics: their specificity of binding, their homogenicity and their ability to produce unlimited quantitities. Antiboidies produced by hybridoma cells are identical and they are powerful reagents for testing for the presence of a desired epitope. The major advantage of hybridoma production is that impure antigens can be used to produce specific antibodies.

Any substitutes that can elicit a humoral response can be used to prepare monoclonal antibodies. Their specificities range from proteins to carbohydrate to nucleic acids. However, monoclonal antibodies are often more time consuming and costly to prepare than polyclonal antibodies and they are not necessarily the best choice for certain immunochemical techniques.

Simplified Procedure

Injecting the antigen to mouse

↓

Spleen cells from mouse cells + Mouse myeloma

↓

Fusion of cells

↓

Only hybridomas grow

↓

Screening hybridomas for desired antibody specificity for ELISA

DNA Vaccines

Most of the commercially available vaccines protect the aquatic organisms against bacterial diseases and are simple inactivated antigen preparations delivered either through immersion or injection with an oil adjuvant. The DNA vaccines which offer several advantages over classical antigen vaccines *viz.*, live attenuated, whole killed and unit vaccines, are being considered a distinct possibility for the future of infectious disease prevention in aquaculture.

Definition

Immunisation of animals with naked nucleic acid encoding antigens under the control of a variety of gene regulatory elements has been described as genetic vaccination or polynucleotide vaccination. This new technology arise from the discovery by Woff and his co-workers in 1990 that injection of pure plasmid DNA (naked DNA) into muscles of mice resulted in long term reporter gene expression in transected muscle. However, there are only a few reports on the use of DNA vaccines of shrimp.

Plasmid Constructs

DNA vaccines most commonly comprise double stranded closed circular plasmid DNA. Plasmid DNA constructs used for gene vaccination are similar to vaccination of therapeutic genes. Basically they have three major units.

1. A plasmas backbone that delivers adjuvant, mitogenic activity and convenient cloning sites for the insertion of genes of interest.
2. A transcription unit comprising strong viral promoter/enhancer sequences, antigen cDNA and polyadenylation sequences which together direct protein synthesis.
3. Markers used to ensure that only plasmids containing bacteria will propagate during culture.

Dose and Volume

Larger volumes are usually associated with a reduction of the reporter gene activity, but less variation between individuals. Typical doses for the shrimp fall in the range of 1.50 ml found to be effective, again it varies according to the size and species of the animals.

Immune Responses to DNA Vaccines

DNA vaccines are known the trigger both non-specific and specific immune systems. In the latter case, they effectively stimulate both humoral and cellular immune responses whereas only the less efficient humoral responses is stimulated by conventional vaccines.

Advantages

DNA vaccines are relatively inexpensive easy to produce and require an identical production process and effective in eliciting both antibody and cell mediated immune responses. Multivalent vaccines can also be easily prepared by mixing together different plasmids.

8. Aquaculture System Management

Recirculation System

Need for Recirculation System

RAS (Recirculatory aquaculture system) can be described as closed or semi-closed culture systems in which mechanical and biological treatment facilities continuous reuse of culture water. RAS in its most advanced form is a production system placed in an insulated building independent of the outside environment with a high degree of water recycling.

Functions of RAS

The aim of RAS is to maintain an acceptable water quality by continuously circulating the water through a treatment system. In a treatment system, bacteria is used to remove organic matter and ammonia. The bacteria also influence water quality by consuming oxygen and by producing CO_2 produced strongly depend on pH of the water. Since the pH controls the equilibrium of those components with less harmful NH_4^+ and HCO_3^-. Dissolved organic matter is not toxic by itself but can cause hygiene problems when it accumulates.

Description of Basic Process

1. Removal of suspended solids
2. Removal of dissolved organic matter and ammonia
3. Removal of waste gases
4. Temperature manipulation
5. Addition of oxygen

Table 1: Comparative Analysis of Conventional and Advanced Recirculation System

Characteristics	*Conventional Recirculation System*	*Shrimp Pond Recirculation System*
(i) Size	Small (raceways)	Large (ponds)
(ii) System components		
(a) Culture system	Tanks	Ponds
(b) Solid removal	Settling and pressurized mechanical filteration	Settling
(c) Aeration	Mechanical	Photosynthesis and mechanical
(d) Nitrificaiton	Inert–Sub components	Photosynthesis and pond surfaces
(e) Denitrification	Require anaerobic sub-component	Ponds mud
(f) U/V, ozonation	Possible	Not likely
(iii) Energy	Intensive	Less intensive
(iv) Solar	Small	Large and important
(v) Water replacement	0–10%	0 to < 5%

Raceway Based Culture System

Several factors have limited the growth and expansion of the pond based shrimp farming industry. Extensive and semi-intensive pond production systems are typically managed with high rates of water exchange and this production method raises environmental problems. In response to above, the recent research is focused on the use of green house enclosed raceways for intensive to super-intensive shrimp production.

Greenhouse–Raceways

Raceways offer several advantages over pond based systems. The raceways can be managed with zero to minimal water exchange thus greatly reducing environmental impacts due to effluent discharge. Biosecurity protocols can be implemented to manage disease vectors. This system can also provide opportunities for inland culture operations and year round production (discussed in detail in another chapter).

Chapter 10
Scope for Genetic Improvements in Shrimp Aquaculture

Sustained production of farmed shrimp requires the use of genetically selected and improved shrimp. Genetic traits of interest include such things as improved survival through disease resistance, improved growth, improved feed conversions, and increased meat yield. These improvements, especially disease resistances, are necessary for further growth of the industry. It would result in increased shrimp production without any further increases in farming areas.

Selective breeding and genetic improvements of shrimp require long-term commitments to closed life-cycle culture, an appropriate selection programme (Cedeno *et al.*, 1999), substantial capital investments, and many years of waiting before profitable returns on investment are realized. Lack of protection for investment is a major impediment for financial commitments by commercial interests. The sale of fertile seed means that people purchasing seed can rear them to broodstock and produce seed for sale in competition with the original investor. Those purchasing improved, fertile seed derive all the benefits of genetic improvements. Triploidy induction is one means of producing sterile seed, but, as we discuss below, mass production of triploid shrimp seed is not yet possible.

Unless means are perfected of inducing sterility in genetically improved marine shrimp seed, we are unlikely to see major improvements in shrimp stocks through genetic manipulations unless these improvements are supported by governments and made available to the industry. Some governments have made commitments in this regard, but as yet we do not see many widespread applications. Most shrimp seed still come from wild-caught broodstock.

A. Closed Life Cycle Culture

Substantial increase in shrimp stock improvements is not possible without widespread use of closed-cycle, captive reproduction of shrimp. Continued reliance on wild–caught broodstock and seed could doom the shrimp culture industry. While only the marginal production increases, they are continuously made susceptible to disease problems. Crop loses create economic instabilities, and in the long run higher shrimp prices.

During the last 10 years, more than 95 per cent of all cultured marine shrimp production consisted of five species: *P. monodon* (tiger shrimp), *P. merguiensis* (banana shrimp), and *P. orientalis* (fleshy shrimp) in Asia; and *L. vannamei* and *P. stylirostris* in the Americas (Csaves, 1994). *Penaeus monodon* accounts for more than 50 per cent of the cultured shrimp production worldwide (Rosenberry, 2000). Closed life-cycle culture of *P. monodon* has been accomplished (Menasveta *et al.*, 1991, 1993, 1994; Sangpradub *et al.*, 1994; Withyachumnarnkul *et al.*, 1999), but is not widely used. Commercial culture of *P. monodon* relies almost entirely on wild-caught broodstock and seed. Closed life-cycle culture of *L. vannamei* and *P. stylirostris* is routinely done in many places such as the mainland U.S., Hawaii, and Tahiti, but the majority of commercial production with these two species still relies on wild-caught broodstock and seed. Nevertheless, we are likely to see more genetic improvements in these latter two species because of their more widespread and reliable closed life-cycle culture.

B. Genetic Selection

When wild stocks of any species are cultivated in closed life-cycle conditions, natural selection still operates even without conscious selection for specific traits (Pullin *et al.*, 1998). When captive-bred populations are also subjected to artificial selection in addition to natural selection, domestication is even more rapid (Doyle, 1983). Domestication leads to reduced genetic diversity, but at the same time produces cultivars that perform better, under specific culture conditions. Domestication, if it proceeds long enough, invariably leads to animals that are unsuitable or unable to survive in the wild.

The vast majority of the world's farmed shrimp production comes from wild-caught stocks that have no selection for any culture traits beyond the species level. Historically, this approach was necessary because life-cycle of penaeid shrimps were not closed until about 1940 (Hudinaga, 1942). It took another 30 years before this basic knowledge resulted in methods for mass producing marine shrimp seed. Today, most shrimp seed still come from the wild or from wild-caught broodstock because they are relatively lower priced. However, there is growing momentum towards greater reliance on hatchery–produced seed from closed life-cycle stocks. The primary motivation for this change now is the need for disease–free and disease–resistant seed. Although that is the primary driving force, closed culture also allows selection, either conscience or unconscience for other desirable culture traits.

(i) Selective Breeding in Shrimps

Selective breeding programs for marine shrimp would, require long-term commitments to shrimp culture for species of interest, especially *L. vannamei* and *P. monodon*. Breeding goals may differ somewhat depending on region, but two traits have universal importance; growth rate and disease resistance (Gjedrem and Fimland, 1995). Although expected growth rate improvements in shrimp through selective breeding programs are unclear at this time, Pullin *et al.*, (1998) indicated, "....rapid and significant positive response to selection for improved growth, " for *L. japonicus* in Australia. Wyban (1992) and Benzie *et al.* (1997) reported indications of genetic variability in growth with *L.vannamei* and *P. monodon*, respectively. These observed variability are favorable indications that growth can be increased through selective breeding since selection for any improvement requires variability in the selected trait.

Detailed studies have been carried out on the effects of selective breeding on fish growth, especially growth of salmon, trout, and tilapia, where increased growth rates of 10 to 23 per cent per generation were found (Gjedrem and Erling, 1995). Substantial yield increases have also occurred with other animal production sector such as hens, pigs, and dairy cattle through conventional genetic selection (Eknath *et al.*, 1991). Economic returns on these investments in genetic selection can be substantial. With Atlantic salmon in Norway, a 10 per cent increase in growth coupled with an 8 per cent decrease in early maturation would result in $20 million/yr economic gain when the industry produced 100,000 MT/yr (Gjerde and Olsen, 1990). Estimated cost for this selection program was $2.5 million/yr. One-year growth increase of 14.6 per cent has been obtained in Norway through selection (Gjerde, 1986).

Selection for disease resistance in marine shrimp has already been achieved in the Western Hemisphere and Polynesia. The earliest selection program began in Tahiti during 1980 (Bedier *et al.*, 1998). During 1980, IFREMER imported *L. vannamei* and *P. stylirostris* from Mexico. Culture performance of *L. vannamei* declined during the 1980s due to IHHNV infection, which resulted in runt deformity syndrome (RDS). At the same time, *P. stylirostris* cultured along with *L. vannamei* gave good culture performance. Tests on *P. stylirostris* confirmed that they carried IHHNV but were not affected by it, because they produced yields of 20 MT/ha/yr. under intensive culture. *P. stylirostris* have now been breed for more than 15 generations in closed life-cycle culture without further imports. During 1993, Tahitian *P. stylirostris* were reintroduced into Mexico and proved superior to local *P. stylirostris* strains during pond growout comparisons. Tahitian *P. stylirostris* also demonstrated reduced susceptibility to TSV. Another captive *P. stylirostris* population in Venezuela has also proved to be IHHNV resistant (Persyn, 1995). These strains now provide excellent SPR (specific pathogen resistant) stocks for wider culture throughout the Americas (Rosenberry, 1999).

(ii) SPF–SPR L. *vannamei* Stock Development

During 1989, six U.S. institutions of the Gulf Coast Research Laboratory Consortium (known as the Shrimp Consortium) began developing SPF and genetically improved *L.vannamei* (Pruder *et al.*, 1995). They also developed a second SPF-TSV-resistant (specific pathogen free) strain of *L. vannamei*. Culture performance of the original SPF strain proved superior to non-SPF shrimp in Hawaii, Texas, and South Carolina with greater growth, survival, and yields with reduced FCR (Beidier *et al.*, 1998). However, large-scale trials in Ecuador proved less impressive, with these SPF shrimp providing poorer performance in many cases compared with non-SPF shrimp. The conclusion was that SPF shrimp perform well in SPF environments, but that most farming areas have little or no means of providing the necessary biosecurity needed during growout. In these areas, SPF shrimp may be more vulnerable, and SPR shrimp would be more appropriate. It was also observed that the original SPF *L.vannamei* performed better well under SPF conditions, and it outperformed the SPF-SPR *L.vannamei* strain with respect to growth, survival, and yields.

(iii) Virus Resistant *P. stylirostris* Stock Development

Although SPR shrimp are not widely used yet, Rosenberry (1999b) reported that at least two domesticated, genetically selected SPR strains of this species (*P. stylirostris*), which are resistant to IHHNV, are currently being developed and marketed in the Americas. In some regions, these SPR stocks of TSV-and IHHNV-resistant *L. stylirostris* are replacing *L. vannamei* stocks in culture. There is also reason to believe that some wild stocks have developed at least partial resistance to certain viruses in areas where these viruses caused earlier, widespread disease problems with cultured stocks (Rosenberry, 1999b; Jory, 1999b).

(iv) SPF and SPR *P.monodon* Production Trials

Domestication and selective breeding of *P. monodon* has just begun (Withyachumnarnkul *et al.*, 1999) in Asia. The ultimate goal is to provide SPF and SPR *P. monodon* seed at competitive prices. Realization of this goal may take perhaps 5 years or more. In the meantime, the strategy for dealing with WSSV and YHV are based on prevention by excluding the viruses and their carriers from culture ponds. The exclusion process begins with assessment of disease status of wild-caught broodstock using sophisticated analytical techniques, including polymerase chain reaction (PCR), molecular *in-situ* hybridization technique, and others (Clifford, 1999). Presently, only PCR provides reliable WSSV detection in broodstock, larvae and PL. Shrimp that test positive through such molecular analysis are eliminated. Farmers typically demands a certificate of assurance from the PL provider before they purchase PL and stock their ponds. Exclusion procedures are also practiced at farms. Farm management practices include (*i*) quick lime (CaO) application to soil between crops to disinfect pond bottoms and to kill disease carriers; (*ii*) disinfecting of pond source waters with chlorine and/or short duration pesticides to kill disease carriers; (*iii*) stocking disease free PL at lower densities to reduce stress; (*iv*) feeding only prepared (sterile) feeds; (*v*) zero water exchange, take steps to prevent its spread to other ponds on the same farm, and/or to other farms in the area. Farmers who practice the shrimp farmer's code of ethics are required to notify farmers in their area who share common source waters of disease outbreaks on their farm. These neighbouring farmers are advised to stop pumping water into their ponds for at least 1 week while the infected farm decontaminates and disinfects its effluent waters. These disease exclusion practices, while not fail-safe, have proven effective. Shrimp production in Thailand increased during 1998/99, at least in part due to these disease management practices.

C. Ploidy Induction

Triploid animals are those with an added set of chromosomes (three sets) rather than the normal two sets (diploid). This chromosome set addition is usually achieved through temperature, pressure, or chemical shock treatments soon after fertilization to force 2nd polar body retention by the fertilized egg (Purdon, 1983). The 2nd polar body is produced during egg meiosis and contains a full chromosome set. However, shock induced triploidy is seldom 100 per cent triploids (3N). Tetraploid animals are produced by shock treatment just before first cell division, following egg fertilization and after the 2nd polar body has been lost (Myers, 1986). At this time, both sets of (2N) chromosomes have replicated (4N) in preparation for first cell division. Shock at this time disrupts cell division and results in an organism with four chromosome sets (4N) in all its cells.

Advantages of Sterile Shrimp Seed Production

The single most important, likely benefit of triploidy with penaeid shrimp culture is seed sterility. Production and sale of sterile seed assures seed producers that their investment is protected from pirating. Without such assurance, a potential investor would be hesitant to make the substantial investment necessary to produce closed life-cycle, genetically improved shrimp stocks and seed. At this point of time its not clear whether all triploid penaeid shrimp species will perform any better or worse than normal diploids in growout culture. Triploid *P. chinensis* grew about 30 per cent larger than diploids in a laboratory setting, while ovarian and egg development were greatly retarded in triploids (Xiang *et al.*, 1999). With fish, triploids often have superior culture performance compared with diploids, including increased growth, lower FCR, and greater percentage yield of edible flesh (Fast *et al.*, 1995; Qin *et al.*, 1999), while tetraploid fish normally do not grow well. Perhaps the greatest value of tetraploid shrimp is their use in crossbreeding with diploids to produce 100 per cent triploids. Culture performance and reproductive competence of tetraploid shrimp are presently unknown.

The produced sterile shrimp seed have other benefits besides protection of investment in stock improvements. The added benefits include reduced risks from escapement of exotic shrimp species, and/or escapement of genetically altered or selected shrimp within their known range (Arthington and Bluhdorn, 1998). In the first case, exotic shrimp that are fertile could potentially become established to the detriment of indigenous shrimp species. In the second case, interbreeding of wild stocks with " improved " culture stocks could potentially result in undesirable, genetic impacts on wild shrimp.

Ploidy induction in penaeid shrimp is more difficult to achieve than with fishes. This difficulty is caused by the fragile nature of shrimp eggs soon after spawning. Newly spawned eggs are encased in a gelatinous layer that takes about 15 min to form into a semirigid membrane around the egg (Hudinaga, 1942). Triploidy induction requires shock treatments during the first few minutes, at which time eggs are very sensitive to any handling. Furthermore, the rigid membrane that forms later is also sensitive to handling before first cell division. Nevertheless, several researchers claim to have produced triploids, mostly using chemical shocks for 2nd polar body retention. Triploidy was reportedly induced in Chinese fleshy prawn (*P. orientalis*) using cytocalacin B (CB) and 6–DMAP chemical shock treatments (Bao *et al.*, 1993; Xiang *et al.*, 1998, 1999), and with kuruma prawn (*P. japonicus*) in Australia using temperature and chemical shocks (Nigel Preston, 1997). Fast *et al.* (1999) had successfully induced tetraploidy in *P. vannamei* using temperature shock. Dumas and Ramos (1999) produced triploid *L. vannamei* by the same means. However, no well-documented cases where triploid or tetraploid penaeid shrimp were cultured to adult size, and where their culture performance was evaluated in commercial pond settings were available. Chances are good that the efficacy of large-scale ploidy induction in penaeids will be fully explored during the next few years. If these efforts prove successful, ploidy induction tactics will simulate greater investments in genetic improvements of shrimp stocks, and they will reduce environmental risks associated with use of exotic or genetically selected or altered shrimp.

D. Transgenic Shrimp

Shrimps that have genes (DNA) permanently inserted and incorporated into their genetic material are called transgenic. Genes are usually inserted at the one cell stage of eggs or sperm just before fertilization, or into the one or two cell stage just after fertilization. Genes are usually inserted using microinjection, retrovirus infection, particle gun bombardment, or electrical techniques (Chen *et al.*, 1995). Growth hormone gene insertions are the most common transgenic transformations with fishes. Often, inserted genes do not become incorporated into chromosomes soon after they are inserted, but remain as extrachromosomal units. If they incorporate into chromosomes during later development as the multicellular stage, this usually results in transgenic mosaics where only some cells have genes integrated into chromosomes. Delayed transgene integration and resultant mosaics are detected by mating these animals with animals that have not been injected with transgenes. If gene insertions were successfully incorporated into atleast one chromosome of all cells, then at least 50 per cent of the resultant crossbred offspring will carry at least one transgene. Stable integration of the transgenes is an absolute requirement for continuous vertical transmission to subsequent generations and establishment of a transgenic fish line (Chen *et al.*, 1995).

Dozens of transgenic plant cultivars have been approved for agricultural uses in the U.S., since 1992 and perhaps hundreds more are under development (Yoon, 1999). Transgenic plants use genes taken from viruses, bacteria, insects, and other animals. During 1999, 20 to 45 per cent of the U.S. production of corn, soybeans and cotton were from transgenic plants. While progress with transgenic plants has been phenomenal during the last 10 years, there has not been a single transgenic animal approved for human consumption by the U.S. Food and Drug Administration till date. Although many transgenic aquatic species have been produced, none have passed all the necessary certifications.

Further transgenic work with crustaceans is still in its infancy (Bachere *et al.*, 1997). To date, most transgenic work with animals has involved nematodes, fruit flies, sea urchins, frogs, laboratory mice, and farm animals (Chen *et al.*, 1995). Transgenic research with marine shrimp is particularly difficult because successful outcomes rely on a number of necessary conditions, many of which are not met, including closed life-cycle culture of shrimp over many generations, basic knowledge about a shrimps genome, basic knowledge about genetic make-up of shrimp disease causing bacteria and viruses, and reliable methods of inserting transgenes and their promoters into shrimp genomes. This work is made even more difficult by the fact that as yet, there are no proven techniques for culturing crustacean cell lines. If these difficulties can eventually be overcome, there are some advantages for producing transgenic shrimp from shrimp strains bred using conventional genetic selection methods. One of the main advantages is that transgenic transformations can be achieved without otherwise altering the shrimp's genetic make-up (Bachere *et al.*, 1997). Conventional genetic selection strategies often result in counter-selection problems and loss of some desirable traits.

In addition to technical problems in producing transgenic shrimp, there are also potential environmental, social, and ethical concerns (Kapuscinski and Hallerman, 1990a; Kohler *et al.*, 1992, Yoon, 1999). Environmental concerns include potential impacts connected with escapement of trangenic shrimp from commercial culture and/or research sites, as discussed above. The American Fisheries Society (Kapuscinski and Hallerman, 1990b) is on record as urging caution with use of all transgenic fishes, and secondly recommends using sterile transgenic fish in commercial aquaculture. The same cautions apply to other aquatic cultivators such as shrimp. In addition to environmental concerns, there is considerable consumer concern about the use of transgenic animals for concern has resulted in public policies against importation and sale of transgenic food items. This policy has already impacted U.S. grain sales to Europe, and should be carefully assessed before large-scale production of transgenic shrimp is undertaken.

Scope for Genetic Improvement Studies in Native Species

Increased, stable production of farmed shrimp ultimately requires large-scale use of domesticated and genetically improved shrimp for culture. Required genetic improvements include disease resistance, increased survival and growth rates, better FCR, and perhaps increased meat yields. Use of disease-resistant shrimp with increase survival and growth would greatly increase world shrimp production from existing farms without any further increases in farming areas. A fundamental requirement for these genetic improvements is use of closed life-cycle culture of shrimp species of interest and selective breeding programs that achieve desired results. Closed life-cycle culture of SPF and SPR (for IHHNV and TSV) is most advanced with *L. vannamei* and *P. stylirostris*. We anticipate greatest near-term advances with these two species, and even more widespread use of hatchery-reared SPF and SPR seed. Similar efforts must be directed on other important species such as *P. monodon, F.indicus,* etc. Genetic selection programs are currently underway for other improvements and should provide measurable benefits within the next few years. Work with transgenic shrimp is still in its infancy and many years away from possible commercial applications. With all genetically selected or altered shrimp, there are serious environmental, social, and ethical concerns that also must be addressed.

Index

A

Agmasoma penaei, 146

Ameson nelsoni, 146

Artemia, 108, 111

Artemia, 35

B

Bacillus, 91

Bioenrichment diets, 158

Biotechnological applications, 151-170

  aquaculture system management, 168

  beta-glucans, 159

  biofilters, 152-153

  biological treatment, 155

    activated sludge system, 155

    aerated lagoons, 156

    aerobic process, 155

    anaerobic treatment, 156

  bioremediation, 151-152

  chlorination, 156-157

  constructed wetlands, 153-154

  defense enhancement by immunostimulants, 162

  definition, 166-167

  description of basic process, 169

  disease management, 163

  disinfections, 156

  DNA vaccines, 166

  dose and volume, 167

eyestalk ablation, 160
feed management, 157
 bioenrichment diets, 158
 microencapsulated diet, 157
 X-linked protein microencapsulzte, 157-158
functions of RAS, 168
greenhouse-raceways, 170
hatchery techniques, 160
immune responses to DNA vaccines, 167-168
immunostimulants, 161-162
incorporation of yeast in aqua feed, 158-159
mannan oligosaccharides, 159
monoclonal antibody, 165
need for recirculation system, 168
nucleotides, 159-160
ozonation, 157
PCR, 163-165
physico-chemical treatments, 156
 coagulation-flocculation, 156
plasmid constructs, 167
 screening and sedimentation, 154-155
 separation of oil and grease and flotation, 155
probiotics, 162-163
raceway based culture system, 169-170
recirculation system, 168
selection of suitable bacteria, 152
simplified procedure, 166
techniques of eyestalk ablation, 160-161
types of biological filter, 153
water quality management, 151
– treatment, 154
 primary treatment, 154

C

Chaetoceros sp., 55
Chlorella sp., 55
Corpus Christi, 27
Cultivable penaeid shrimp species, 37-43
 Fenneropenaeus indicus, 40
 Litopenaeus vannamei, 42
 Penaeus monodon, 38
 Penaeus semisulcatus, 42
Cynadon plectostachus, 60
Cynadon, 60

D

Disease and health management, 131-150
 bacterial diseases, 142
 bacterial shell diseases, 143
 bacterial white spot syndrome, 144
 baculovirus penaei, 136

BMN and other type C baculoviruses, 137
chlamydial diseases, 142
comfirmatory tests, 138
daily observations, 132
disease caused by epicommensals, 147
filamentous fouling and gill disease, 147
fungal diseases, 145
fusariosis, 145
gill associated virus, 142
gregarines, 147
haplosporidians, 147
hepatopancreatic parvo like virus, 133
infectious hypodermal and haematopoietic necrosis virus, 134
larval mycosis, 145
lymphoidal parvo like virus, 135
micro-sporidians, 146
monodon, 136-137
mycobacterium infections, 145
parastitic disease, 146
pathogen detection, 132
REO like viruses, 136
rickettsial diseases, 142
taura syndrome, 141
toga viruses, 136
vibriosis, 143
viral diseases, 133
white spot syndrome, 138
yellow head disease, 139

F

F. solani, 145
Feed management, 79-90
Fenneropenaeus indicus, 38
Fisheries College and Research Institute, 27-28
Fungal diseases, 145
Fusariosis, 145

G

Gill associated virus, 142
Gregarines, 147

L

Lab-lab, 61, 62
Lactobacillus acidophilus, 92
Lactobacillus, 91
Larval mycosis, 145
Litopenaeus vannamei, 38

N

Nitrogen compounds, 52
Non-infectious diseases, 148
ascorbic acid deficiency syndrome, 148
chronic soft shell syndrome, 148
cramped muscle syndrome, 148
pale coloration or blue diseases, 149

P

Penaeus chinensis, 38
P. japonicus, 38, 47, 55, 89

P. merguiensis, 38, 46, 61
P. monodon, 37, 38, 46, 48, 61
P. schmitti, 47
P. semisulcatus, 38, 46
Probiotics and immunostimulants in shrimp aquaculture, 91-103
careful experimental designs and analysis, 99-103
commercial immunostimulants, 98-99
glucans, 97-98
immunostimulants, 95-97
probiotics as pond water additives, 93
for shrimp larvae, 93-95
yeast, 97
Protein microencapsulzte, 157-158
PVC, 35, 36

R

Reservoir pond, 27

S

Salmonella, 85
Scope for genetic improvements in shrimp aquaculture, 171-184
clay content acidity, 5-7
closed life cycle culture, 172-173
genetic selection, 173-174
ploidy induction, 178-179, 180-181
scope for genetic improvement studies in native species, 183-184
selective breeding in shrimps, 174-176
SPF and SPR *P. monodon* production trials, 177-178, 179-180
SPF-SPR *L. vannamei* stock development, 176
transgenic shrimp, 181-183
virus resistant *P. stylirostris* stock development, 176-177
Shrimp culture systems, 1-36
biological factors, 10
predators, competitors and burrowing organisms, 10
seed resources, 10
vegetation, 10-11
wood boring organisms, 10
construction in areas with acid sulfate soils, 7-9
criteria for site selection, 1-2
design and construction, 14
earthen and lined earthen ponds, 1-2
earthen ponds, 14
ecological factors, 2
currents prevailing in the immediate water source area, 3-4

depth of topsoil and characteristics of subsoil, 9
evaporation rates, 4
load bearing capacity, 9-10
percolation rate, 9
pollution, 4
rainfall, 4
salinity, 2
soil, 4-5
tidal characteristics, 2-3
water quality, 2
general layout, 14-15
canals, 19-20
control of erosion and leakage, 18-19
dike formation, 17-18
water control gates, 20-22
– distribution system, 15-16
greenhouse structure, 28-30
HDPE sheet/synthetic fibre spread, 23
LDPE sheet, 23
lined ponds, 22-23
lining materials, 23
number of soil samples, 5
potential acidity, 5-7
raceway ponds, 24
air diffusers, 36
– lift pumps, 35
backup generator, 36
blowers and air distribution system, 35
bottom manifold system of raceway tank, 34-35
design and construction of an intensive nursery, 24-27
lighting facility, 36
lines and valve layout for raceway, 30-31
multiport valves, 33
oxygen injection, 33-34
pipe, 30-31
pump, 30-31
raceway outlets and filter peoples, 32
rapid and filters, 32
temperature control, 36
selection of lining materials, 24
site selection for raceway ponds, 13-14
social and economic factors, 11
accessibility, 11
availability of labour, 12
– and cost of inputs, 12
– of technical assistance, 13
land cost, 11
marketing outlets and prices, 12-13

possible legal and institutional constraints, 13
users competing for land and nearby waters, 13
system description, 27
testing of clay content, 5
Shrimp nutrition and feed management, 79-90
antibiotics, 85
anti-nutritional factors, 83
biogenic amines, 82
carotenoids, 89
fat soluble vitamins, 84
feed formulation, 80
horizontal transmission of pathogens, 85-86
imbalances in feed formulation, 81
ingredient selection and quality criteria for feed formulation, 81-82
micronutrients in enhancing immunity, 86
mycotoxins, 82-83
toxicity of additives, 83-84
trace minerals, 84-85
trace minerals, 89-90
vitamin A, 88
vitamin C, 87-88
vitamin E, 88
Shrimp raceways, 104-130
acclimation process of post-larvae, 106-108
algal cell counts, 113-114
ammonia, 114-115
bottom condition evaluation, 117-118
counting and storage, 105
daily assessment of population, 119
disease assessment, 119-120
dissolved oxygen, 111-112
feed and feed ration, 108-110
fertilization, 105
filling, 104-105
filter screen size, 117
harvest, 121
harvesting and data records, 122-123
hauling/harvest tank preparation, 121-122
population sampling, 123
raceway preparation, 122
transportation of PL, 123-124
weighing station, 122
health management, 121
hydrogen sulphide, 116
managing water quality parameters, 111
monitoring of population growth, 120

nitrate, 115-116
nitrite, 115
pH, 114
preacclimation, 105-106
preparation of raceways for stocking, 104
salinity, 112
sample size and sample preparation, 119
seechi disc visibility, 112
supplemental feeding, 110-111
temperature, 111
troubleshooting in raceways, 124-130
water exchange, 116-117

Stocking management, 57-78
benthic animals, 67
estimation of the desired fertilizer application rate, 67-70
holding and growing small post-larvae to a larger size for stocking, 75-76
identifying shrimp larvae, 72-73
lab-lab, 61-63
method of growing and managing
lab-lab, 62-63
phytoplankton, 64-67
mode of fertilizer application, 70-71
nursery ponds, 76-77
phytoplankton, 62-64
pond preparation, 57
biofeed production in ponds, 61
drying the pond bottom, 57-59
improving or controlling soil acidity, 59
liming, 59-61
poisoning predators and pests, 61
provisional key for the identification of post-larvae, 73-74
recommendations for managing natural food organisms, 71
transferring fry from nursery to growout ponds, 77-78
transportation of shrimp seeds, 74-75

Streptococcus, 91

T

Texas A & M University System, 27

Texas Agricultural Experiment Station, 27

Thelohania duorara, 146

Toxic disease syndromes, 149-150
aflatoxicosis, 149
black gill disease, 149-150
gas bubble disease, 150
muscle necrosis, 150

V

Vibrio alginolyticus, 92, 94
V. anguillarum, 92
V. harveyi, 95
V. parahaemolyticus, 92

W

Water quality management, 44-56
    hydrogen sulphide, 55-56
    nitrogen compounds, 52
        ammonia, 52-55
        nitrate, 52
        nitrite, 52
    oxygen, 47-51
    pH, 51-52
    salinity, 46-47
    temperature, 44-45